WORLD WAR II TROOP TYPE PARACHUTES
Allies: U.S., Britain, Russia

WORLD WAR II TROOP TYPE PARACHUTES

Allies: U.S., Britain, Russia

An Illustrated Study

Guy Richards

Schiffer Military History
Atglen, PA

Book design by Robert Biondi.

Library of Congress Catalog Number: 2002116807.

Printed in China.
ISBN: 0-7643-1781-4

Published by Schiffer Publishing Ltd.
4880 Lower Valley Road
Atglen, PA 19310
Phone: (610) 593-1777
FAX: (610) 593-2002
E-mail: Info@schifferbooks.com.
Visit our web site at: www.schifferbooks.com
Please write for a free catalog.
This book may be purchased from the publisher.
Please include $3.95 postage.
Try your bookstore first.

In Europe, Schiffer books are distributed by:
Bushwood Books
6 Marksbury Ave.
Kew Gardens
Surrey TW9 4JF
England
Phone: 44 (0)20 8392-8585
FAX: 44 (0)20 8392-9876
E-mail: Bushwd@aol.com.
Free postage in the UK. Europe: air mail at cost.
Try your bookstore first.

Contents

CHAPTER ONE

Russia

The Soviets were in attendance at the famous demonstration at Brooks Field, Texas on September 28, 1928, when eighteen men dropped by parachute from nine DeHavillands and three Douglas transports and in minutes had assembled three Browning water cooled machine guns for firing.

The U.S. brass ignored it, but the Russians took notes and began developing their own airborne forces. There are subtle hints that the Soviets were developing or considering a parachute arm during the period 1927-1929, and that the Brooks Field show provided further stimulus. Coincidentally, the first free fall jump in Russia was made in 1926 by the American Lyman Ford, who later became president of Pioneer Parachute Co. He was probably demonstrating an Irvin product.[1]

The 1920s and 1930s were a time of great military experimentation in the Soviet Union. Marshal Tukhashevsky was the main promoter and August 2, 1930 is considered the birth date of the Soviet Airborne when twelve troops made the inaugural drop.[2] OSOAVIAKHIM (Society for the Promotion of Aviation and Chemical Defense) was charged with organizing parachute jumping instruction throughout the Soviet Union. They did a superb job, and by 1936, there were 559 jump towers in 115 training sites.[3]

In 1933 the Soviet Army dropped a light company of sixty-two. By 1935, they were delivering a brigade from the sky. The Russians are therefore the first to capitalize on the airborne concept of delivering units to a prospective battlefield via parachute. In 1941, the Soviets had no less than fifteen parachute brigades.[4]

At this point, their parachute was identical to the Irvin dual training with its non-detachable reserve (U.S. T1, T2), although the reserve had a center pull ripcord handle. Domestic parachute production began in April 1930.[5] Their very first batch of parachutes had been purchased directly from Irvin in 1927 for $600 each.[6]

The standard main parachutes during the war were the PT1A, PD 6, PD 6R[7], PD 36, and PD 41[8] There were additional models: PD 2, PD 7, PD 8, PD 10, and PD 10U[9]. The flat circular canopies had a diameter of 8.76 meters, an area of 60.3 square meters, with twenty-eight gores of four panels each, bias construction with the lowest panels made of silk and the others of cotton percale.[10] The suspension lines were 6.54-6.65 meters long, and were made of cotton with a 125 kilogram tensile strength. The ripcord was 1.07 meters long, held three pins and had a trapezoidal handle that was mounted in a pocket on the left side (left inboard pull).

The reserve canopy had twenty-four gores and an area of 42.5 square meters, which would make it about a twenty-four foot diameter. The suspension lines were 4.5 meters long, and rated at 150 kilograms.[11]

The Soviets began using the static line method and the main pilot chute was removed for this purpose.

This transition was accomplished to reduce losses from jumpers who failed to pull their ripcords.[12] The static line was a flax rope 12-14mm in diameter, 3.9 meters long (about twelve feet, ten inches).[13] The static line was stowed on the right side of the container under the top and bottom pack opening bungee bands. The static line had its own log book that stayed with it in a pocket. Apparently the static line connected to the canopy apex by break cord. By 1942, at least some Soviet parachutes were bag deployed.

Soviet paratroops demonstrate their technique. They exit from a dorsal hatch and a side door, tumble off into space and manually activate their parachutes (from 1935 film footage). Multiple ship drops were coordinated by flag signal from a despatcher in the nose of the aircraft. The Russians pioneered in delivering units via parachute and considering the amount of real estate they had to protect, it makes perfect sense that they would.

Boarding the bomber. The side caps would be replaced by the rib-padded helmet in flight. The main packs are bulky because of the cotton/silk construction (undated, but probably a pre-war photo).

The square main canopy, the PD 41[14], that was fielded then had an area of seventy square meters (677 sq. ft.), and sixteen assistor pockets[15] along the skirt as it was reputedly slow to develop. It could be either static line or ripcord activated. This square form canopy was simple to manufacture and had a built-in forward drive by the omission of two center rear suspension lines, which created a vent. It was standard practice during the 1930s for Soviet jumpers to activate their reserve and descend under both canopies. This was done to reduce oscillation and would also nullify any advantage a steerable canopy affords. The reserve was the PZ 41 with a square cotton canopy of 515 square feet. The PD 41A was silk, the PD 41I was cotton with areas of 750 square feet.[16]

A well known 1935 film shows Soviet paratroops tumbling out of a Tupelov TB-3 bomber from side and top hatches, falling free and activating their parachutes. There appears to have been no technique to the exit, that is judging by the many who slid down the wing, just "get outside and yank the handle!"

In 1941, the Soviets began licensed production of the Douglas DC-3 transports (the PS 84) which facilitated more efficient paratroop delivery. They so admired the "Gooney Bird" that they copied it exactly – the Lisunov 2 added a dorsal gun turret and a wood burning stove.

The Soviets conducted at least fifty[17] airborne operations during the war generally in support of ground offensives, diversions and partisans. The most notable were the drops surrounding Vyaz'ma in January-February 1942. These committed 14,000 troops in the longest continuous parachute operation.[18] The old story asserting that the Soviets dropped troops without parachutes may have a basis in fact. In early 1943 the Germans concluded that Soviet soldiers were deposited in their midst near Vyaz'ma into snow drifts from ground skimming planes.[19]

The Soviet paratrooper emblem during World War II was a royal blue cloth lozenge with a gold parachute symbol between two stylized aircraft. The patch has a recessed red border and a red star on the parachute symbol. The qualification badge was an enameled metal parachute symbol with a gold airplane figure in the middle and a red star at the top.

Soviet World War II paratroop insignia.

Above left and above: PD 41 square canopy. This jumper has landed under both canopies – a practice that ended in 1954.

Left: Russian parachutist under a PD 41 square canopy throws out his reserve. Note that the rear is vented by the omission of some suspension lines.

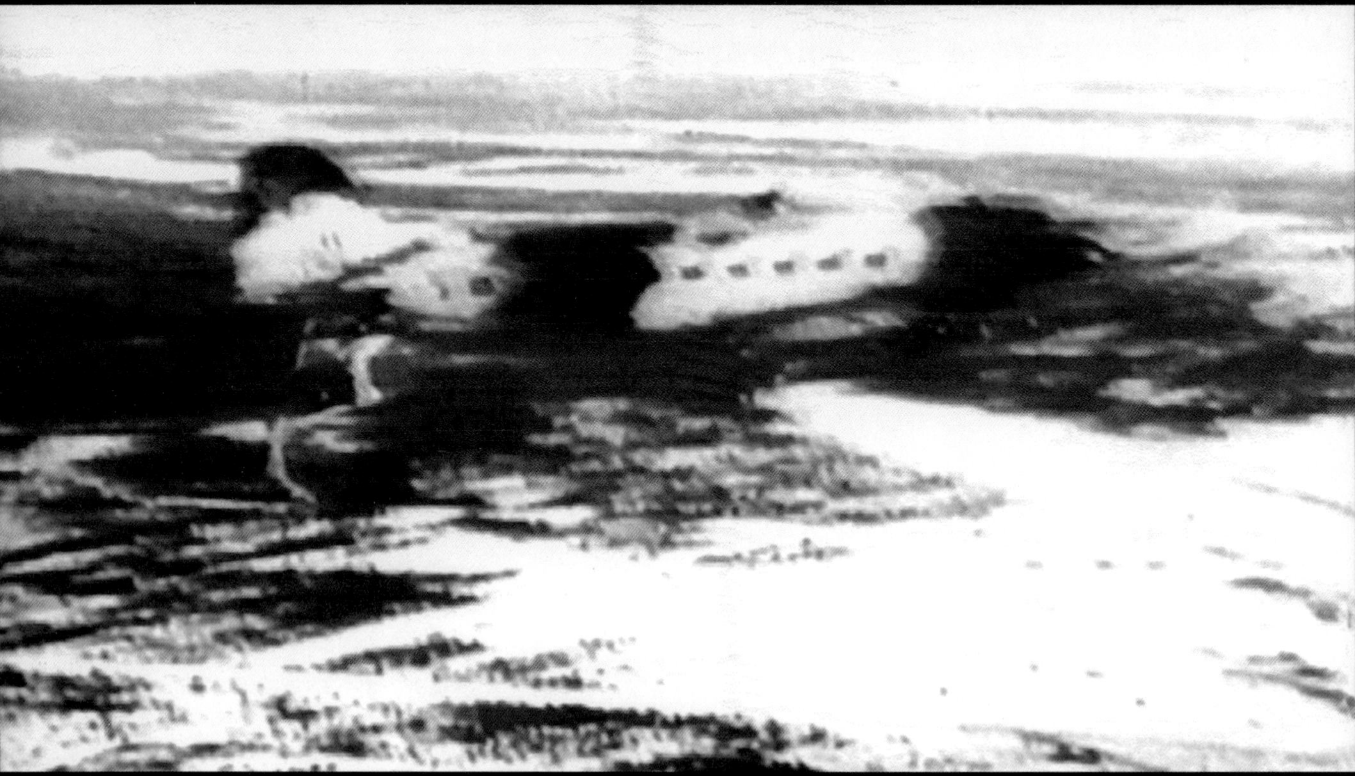

Above: This frame, though difficult to see, shows a canopy being drawn in from a deployment bag. Type unknown. Below: Closer view of the Lisunov 2 as two jumpers pile out. Static lines trail from the open door and a d-bag is between the two.

Russian jump training tower.

CHAPTER TWO

France

In 1935, the French Air Force sent a detachment to Russia for a feasibility study of military parachuting. This detachment became jump qualified in the Soviet technique and returned to France with a recommendation to proceed with developing airborne forces.

Thus the French paras initially belonged to the Air Force. A school was established at Avignon in October 1935.[20] The parachute equipment appears identical to the U.S. T2/T3. This equipment was referred to as "Irving-Russe". They quickly adopted a system with a detachable chest reserve. The method of jumping was to clear the airplane and pull the ripcord. Judging from a photo (admittedly not the most accurate means) the drop altitude seems to be a bit high, maybe as high as 4000'. Like the Russians, they activated the reserve and descended under two canopies.

By January 1938, two companies, the 601 and 602 Air Infantry Groups[21] were trained and ready. Two schools existed, one at Rheims and another at Algiers. Following the armistice of 1940, the small airborne arm was disbanded on July 27, 1940.

Training resumed in unoccupied France despite German protests and two operational jumps were made. On March 15, 1941, five troops were dropped near Vannes to attack an airfield, and on 14 May a company raided a power station near Bordeaux. These operations were strictly intramural affairs and directed at the German controlled (Vichy) regime.[22]

A battalion of paras was formed at Fez in January 1943 under U.S. auspices. This Free French unit was expanded to regimental size by May. It never made a combat drop, but did participate in the invasion of Southern France. French paratroops were also organized in England in 1940 and eventually became part of the SAS brigade.[23]

France can boast the most experienced airborne forces in the world with more combat drops than anyone.

These were made in Indochina 1946-1954 where typically company and battalion formations were deployed by air. France's Paras have made at least 191 combat drops from 1941-1978.

The pre-war paratroop badge or brevet is a parachute symbol holding a gold star across which is the stylized wing of the Air Force pilot's badge.[24]

French parachutist descending under main and reserve canopies. The detachable reserve is connected to the harness via short risers – its container is suspended above the jumper.

This Frenchman has just thrown out his reserve canopy and it is slowly wafting open.

The French were quick to make the reserve a detachable item. The harness here has integral reserve risers. Note left side inboard main ripcord handle.

CHAPTER THREE

Poland

Poland began their airborne arm in 1936 and by mid 1939 had a platoon sized element trained and ready. A second class was under instruction when the war erupted.

The Polish paratroop inception was patterned after the French model, which followed the Russian lead. All used Irvin designed equipment. The Poles began experimenting in 1927 with British made Irvin training types. Licensed production of Irvin models commenced in 1929.[25] (The Poles can claim to be among the earliest pioneers in parachuting, as it was a Polish gentleman, Jodacki Kuparento, who made the first emergency descent via parachute from a burning balloon on July 24, 1808).[26] These systems were identical to the U.S. T3. The earliest jumps were the "pull off" method, which was universal then, and as the name implies had the jumper activate his parachute while clutching the strut of a biplane. The main canopy had a surface area of fifty-seven square meters, which equals about twenty-eight feet in diameter, with a nineteen inch apex vent and twenty-two foot suspension lines. The reserve was twenty-two feet in diameter for an area of thirty-five meters square and twenty gores.

It was standard practice to activate both parachutes[27] in order to reduce oscillation. A jump school was established at Bygoszcz in 1939.

Many of these troops escaped to Britain where their experience was welcome in the formation of the parachute school there. One contribution that reduced landing injuries was the canopy manipulation now known as the "slip". This was accomplished by pulling down on one or two risers, spilling air from the tilted canopy, reducing oscillation and against the wind, and lateral drift. This maneuver is credited to a Pole, Julian Gebolys, and was adopted into training British troops.[28] A side benefit was that slipping placed one in the preferred position for landing.

The other notable Polish technique was their method of dropping through the aperture. This they did in a squat, presenting a smaller surface to the slipstream and presumably lessening the possibility of tumbling.

The Sosabowski Brigade used British "X" type equipment in the war.

Early Polish jump training.

The Polish paras followed the Russian and French practice of descending under both canopies.

Polish parachutists 'chuting up. Their rig is the same as the U.S.T3.

A stick of jumpers waiting to enplane. A crew member leads the squad, perhaps a static jumpmaster.

Right: Polish jump training tower.

CHAPTER FOUR

Britain

The British had the best troop type parachute during World War II. This, the "X" type, was bag deployed with a conventional Smith (Irvin) harness that suspended the jumper from the shoulders. A single point quick release box joined the leg and chest straps, and as no reserve was provided, this made it easy for the trooper to get out of on the drop zone.

The X-type was the product of extraordinary cooperation between the GQ firm and Irving Air Chutes (UK).[29] The GQ concern began in 1932 when James Gregory and Raymond Quilter formed a partnership in competition with Irvin to produce aircrew escape systems for the Royal Air Force. In 1936 GQ offered a troop type static line model to the War Office, but they were disinterested.[30]

The war brought the issue back to life especially after the stunning successes of the Germans' displays of vertical envelopment. Prime Minister Churchill precipitated action in June 1940 with a memorandum stating, "we ought to have a corps of at least 5000 parachute troops." Number 2 Commando was designated the test formation for the concept and posted to Ringway Airfield, Manchester. Training began July 9, 1940 on the ground with a borrowed supply of Irvin training types. These did not have chest reserves. Jumps commenced on July 13 from the tail of Whitley bombers, where the gun turret was removed.[31]

Reportedly, the Irvin parachutes were jury rigged with a static line. This meant that the container was forcibly opened and the canopy flung into the relative wind. On the 25th, after fifty-seven jumps[32], a failure resulted in a trainee's death which suspended jump operations until the cause could be found[33]. Mr. Quilter was able to present his bag-deployed 'Statichute', which was adopted as the X- type. Some problems remained, but by 1944 the X was free of annoying vices and remains in service today with a larger (thirty-two foot) canopy as the PX.

The X-type was made so reliable that the British did not adopt a reserve until 1952. In November 1956, when the Paras were tasked with the drop on the Suez Canal, they declined reserves and packed more ammo instead.[34] Initially, like most, X-type canopies were made of silk. When the supply (Japan was the principal source) became dubious, canopies were fabricated of a cotton weave called Ramex.[35] Silk stocks were diverted exclusively for production of aircrew escape type canopies such as seat and QAC types. Finally nylon became available and standard in 1944. Some canopies had been made of nylon as far back as 1940.[36] X-type harnesses were initially made of undyed cotton with centerline stitching color coded by manufacturer. The hardware was forged stainless steel which was later chrome plated. The quick release box, patented by the Irvin firm in Britain in 1929[37], may release all four lugs and remain secured to the harness by its own strap, or may be retained by a captured lug on the upper left side. A safety locking fork under the release button was not provided. The container had a semi rigid base and the side flaps do not fully enclose the deployment bag giving a packed X-type a less than neat appearance.

The X-type harness webbing is a cotton/linen weave 1 3/4" wide with a tensile strength of 4500 lbs. The harness does not have web loops like the T7, through which the leg straps pass. The leg straps are routed around the main lift web and held fast against the diagonal back strap. The twelve foot static line terminates in a V ring. The static line is 1 3/4" wide, like the U.S. type VIII webbing, but double thickness.

The V-ring snaps onto a variable length strap, which is fixed by a D-ring to the anchor line cable or a strong point inside the jump aircraft. The snap hook was provided with a locking pin after November 1940. A freak fatal accident compelled this modification.[38]

As aircraft were in short supply during the summer of 1940, the RAF pressed obsolete bombers into service. These aircraft were adapted for jumping by installing a portal – the "aperture" – in the belly. Aircraft numbers never exceeded twelve Whitley bombers at the school at Ringway airport.[39] The static line V-ring was snapped onto a web extension called a strop. This arrangement furnished the versatility required for the variety of aircraft used in jump operations. The strop varied in length depending on which airplane was jumped: 6' for the Albemarle, 9'9" for the Dakota (C-47) and 16' for the Halifax.[40]

Dakotas could be rigged with anchor cables in two ways. The cable could run along the floor on the right side or diagonally from the left side of the forward bulkhead to the lower right rear of the cabin. This would cause the static lines to trail out from the middle or bottom of the troop door. The configuration of cable and strop came after considerable experimentation and was necessary to prevent deployment bags from fouling control surfaces. It also kept the developing canopy flow away from the tail cone, tail wheel, and horizontal stabilizer.

The Whitley and Wellington strop arrangements were complex. Whitley jump position, in relation to the aperture:

#1 starboard forward 5'6" strop length
#2 port aft 3'3"

British and Polish paratroops with their weapons in a pre-jump formation. The semi-rigid construction of the X type pack tray is clearly seen. The placement of the static line identifies these as Mark I versions of the X type.

#3 port forward 5'6"
#4 starboard aft 3'3"
#5 starboard forward 6'9"
#6 port aft 4'0"
#7 port forward 6'9"
#8 starboard aft 4'0"
#9 starboard forward 7'10"
#10 port aft 4'9"

The Wellington required a similar staggering arrangement with strop lengths of 3', 5'6", 6'9" and 7'10" depending on jump position. Jumps were made to distribute weight loss so that the aircraft center-of-gravity would not be disturbed. It is no wonder then that the British employed a static jumpmaster, the Despatcher, someone whose job was to keep all of this straight. The Wellington, Whitley, Albemarle and Halifax could all carry ten troops. The Whitley carried the Paras five on each side of the aperture. The Wellington had six fore and four aft. In the Halifax, eight were forward and two aft. Only the Albemarle had all ten forward of its spacious elongated exit slot.

The Whitley aperture was small, little over 3' in diameter (47" at top, 38" at bottom and 24" deep), and all too frequently induced scrapes, bruises and broken noses. Slamming one's face into the cylindrical wall by exiting vigorously was called "ringing the bell". The aperture exit was unpopular.

Jump commands were brief: "Action stations!"

"Green on - GO!"

The X-type Mark I can be identified by the vertical stowing of the static line on the back center of the container[41] Mark designation can refer to the container (outer bag) or deployment bag (inner bag). In this case, Mark refers to the container.

The X-type Mark II repositioned the static line stowing it inside the container beneath the deployment bag at the bottom of the pack tray under four web loops. The static line was passed out through a sewn square grommet on the top container flap. The static line would be deployed vertically from a point behind the jumpers' neck and would be so rigidly controlled that it would have little chance to whip around inducing mischief such as twists, flips, somersaults when dropping though the aperture of a converted bomber.[42]

The Paras dropped from converted bombers through a hole in the belly and the exit attitude was the position of attention. "Go out like a Guardsman!", they used to say, hands grasping their trouser seams.[43]

Right: X Type with Mark I container.

Publicity photo ostensibly showing troops managing canopies on the ground. They have not just landed, as there are no main containers.

"Jumping was a piece of cake!", was another adage and indeed the X-type being bag deployed would have been a painless albeit thrilling experience unlike the bruising punishment of the U.S. T5.

The British then issued no reserve, so they took great care to examine all the details that ensured reliability. Considerable attention was paid to how the suspension lines would be stowed, the size of the stow loops, what they should be made of and how arranged. It was demonstrated by motion picture photography that constant tension must be applied to he deploying lines to prevent snarls, tangles and ultimately malfunctions.[44]

The design of the deployment bag received close scrutiny. Initially there was the Mark I with which the suspension lines were stowed under ten elastic (later web) loops on a flap hat was an extension of the bottom side of the bag envelope. The canopy was retained by a circular break cord tie drawing four beckets, and two closing flaps over the bag mouth.

The Mark II moved the line stowage to the top of the bag with the addition of a flap that covered the first length of lines that locked the bag shut. This precluded lines from rubbing lines as they deployed and prevented searing, melting, fusing and breaking.

The X-type canopy was nominal twenty-eight feet in diameter. Laid out flat they are only twenty-five feet, the difference is attributed to stretch when placed under load. The apex vent is 22" and the suspension lines are 22'6" long. Canopies were bias constructed, at first white, but later dyed a tactical light brown. Some were made from a speckled camouflage pattern fabric. Thirty-two foot cotton canopies were produced, starting in 1941, for dropping agents into Europe.[45] A flexible steel cable extension was adopted for use where only one or two men were to be dropped from bombers. This would give plenty of tail clearance. The technique was adapted for special operations and had one notable failure. During the insertion of OSS detachment 'Union II', on August 1, 1944, USMC Sgt. Charles Perry's cable probably kinked, then severed. Anyway, he had a total malfunction and died on impact.[46] The cotton Ramex constructed canopies were of two fabrics with differing permeabilities. The top panel (D) would be less permeable in order to trap air faster at the crown.[47] These panels are darker than the rest of the canopy. The PX canopy, adopted in the 1960s reversed the porosity layout with the more permeable fabric at the top to reduce oscillation.[48] Another minor but thoughtful improvement was included in the D-bag. This was a protective sock that was 11" long, 4 1/2 " wide at bottom and tapering to 2" at the top that ensheathed the apex. The break cord tie was within this sock and was two turns of fifty lb. nylon cord.

One of the most important innovations was RAF Lt. John. C. Kilkenny's landing method.[49] Previously, parachutists had almost universally been taught to land and perform a forward somersault. Dangerous, because done incorrectly (over the head) can result in a broken neck. The Germans spent many hours drilling their trainees to do a forward roll, ducking the head, and applying one's back to the ground. U.S. troops were constantly reminded, "Don't forget to tumble!"

It is just not possible to execute these school-solution rolls under a drifting canopy, in variable winds, and especially with a full combat load on a night darkened DZ. The trooper merely crashes. The Parachute Landing Fall is the best and most workable method of absorbing the stop.[50] The U.S. Army quietly adopted the PLF in 1943.

Initially British jump school was six weeks duration beginning with two weeks of PT, medical tests, mental assessment and pertinent lectures. One week was devoted to intense combat training. One week of ground instruction e.g. mock door, landing trainers, PLFs. Two weeks of jumping that included one or two balloon drops (one at night) and five from aircraft.[51]

The U.S. Army considered adopting the "X" Type or at least the bag deployment feature in mid 1941. Capt. Yarborough and Lt. Ryder just returned from Britain managed to obtain a small number of Statichutes, which were test jumped. Opinions were divided and the deciding factor came from the U.S. Forest Service, which had just started their smokejumper operations using the canopy-first deployed Eagle system. They allowed that canopy first was faster developing and gave them no problems. So the U.S. Army Airborne went with the T5.

"A" Type

There was another British troop parachute that was intended solely for use by saboteurs, agents and intelligence operators for infiltration purposes. This was the "A" Type. This was a fixed station design functionally the same as a World War I balloonist rig in which the packed canopy was a separate bundle attached to a strong point inside the aircraft. The "A" Type was bag deployed just like the Mark I "X" Type but a small cargo bundle was positioned beneath the risers and above the parachutist's head. This arrangement preceded the leg-bag method of carrying cargo. To quote the official document:

> "The supplies pack is attached at each side to a snap hook fitted to each lift strap, so as to be suspended between the parachute rigging lines and the parachutist during descent."

The document does not state what the maximum permitted weight was, but the most likely the stowed items would be some small mission essential gear like a radio set or beacon.

Canopies were of two sizes, twenty-eight or thirty-two feet in diameter with two foot apex vents.

The systems' operation was simple, the jumper leaped into the aperture or open bomb bay and the risers (lift straps) deployed which extended a cord that withdrew a pin and released the deployment bag to follow the jumper out to the full extent of the static line. This placed the packed parachute out into the slipstream just like the "X" Type. A break cord tie at the mouth of the bag severed by the jumper's weight and allowed the supply bundle and suspension lines to deploy. Then the canopy streamed out and developed.

This system was used routinely on insertions into Norway, the Netherlands and other places. It was a hazardous device to use and killed two agents on missions and injured numerous others during training. Doubtlessly, it made every user rather nervous and hoping for some wind on the DZ to blow the bundle over on landing.

This page: Jump training.

FIG. 2. PHASES OF EXIT (ATTENTION POSITION)

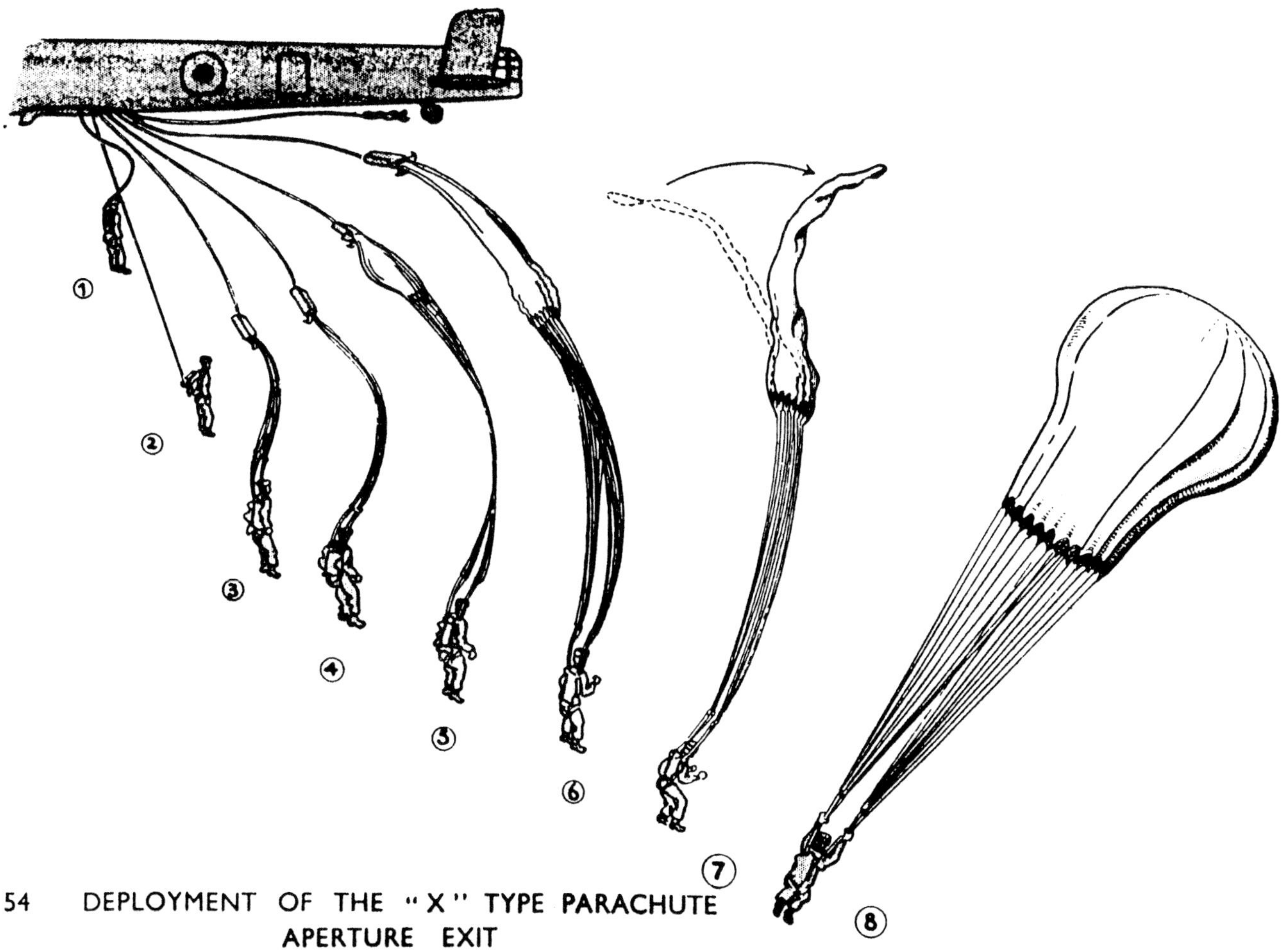

FIG. 54 DEPLOYMENT OF THE "X" TYPE PARACHUTE APERTURE EXIT

Compare with Fig. 55. Note that rigging lines are fully withdrawn before canopy appears.

The slipstream often induced tumbling.

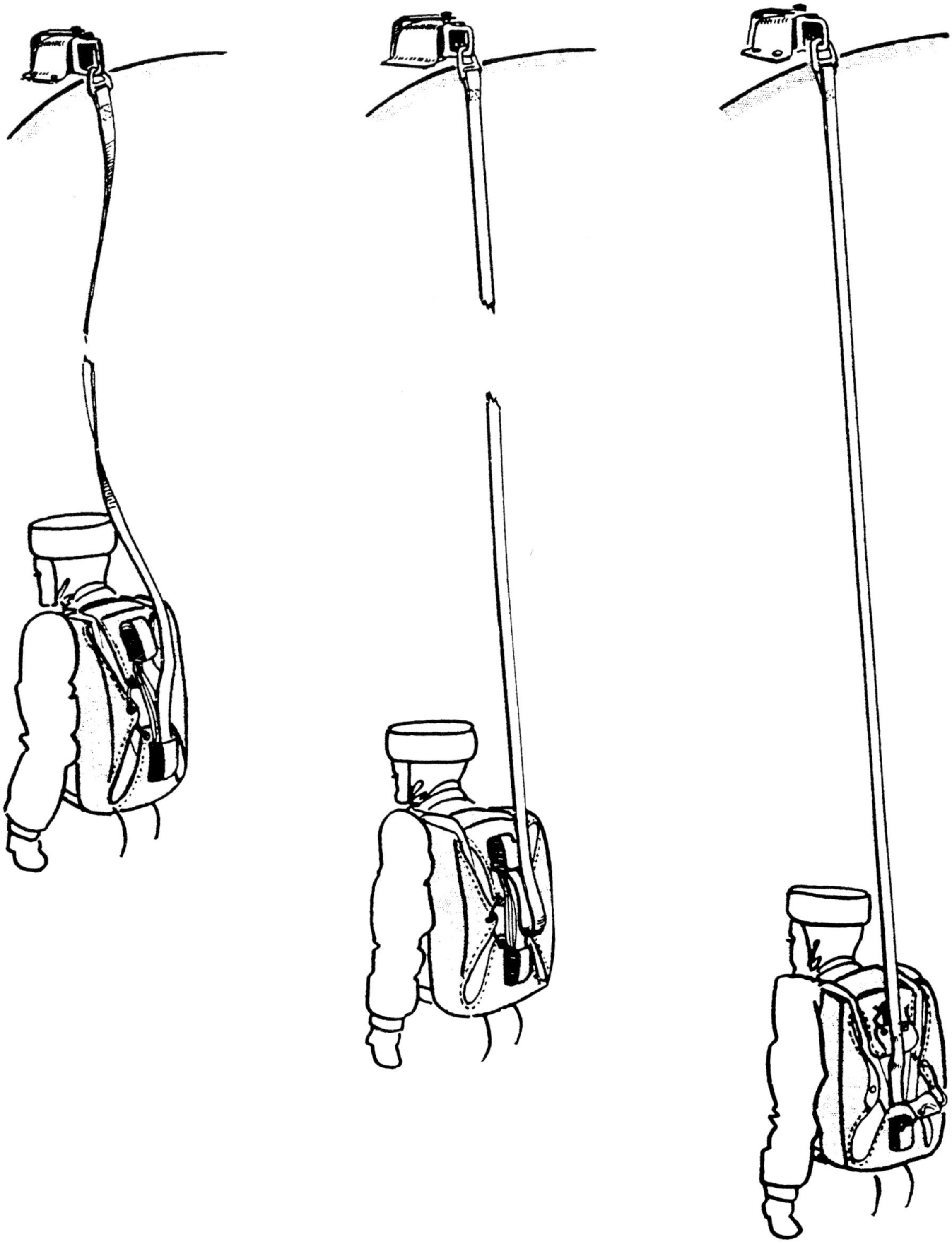

FIG. 46 AUTOMATIC OPERATION OF PARACHUTE

Shows static line being withdrawn from pack by weight of parachutist. The lower end of the static line is tied to the apex of the canopy.

The X type was also assembled with a speckled pattern camouflage. This version appears to have been short-lived and approval was granted to the solid color similar to russet brown.

This formation of British and Polish paratroops stands ready to enplane wearing their X type Mark Is. The bundle worn on the chest is a 'respirator'. They wear the early smocks and rubber helmets and are armed with Bren and 1928 Thompson guns.

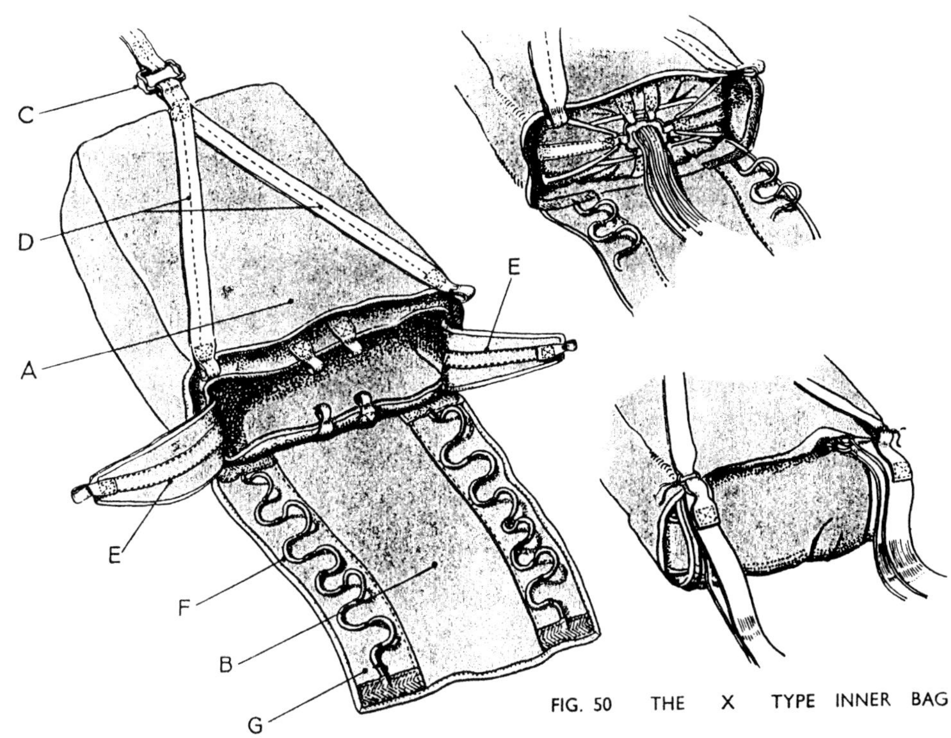

FIG. 50 THE X TYPE INNER BAG

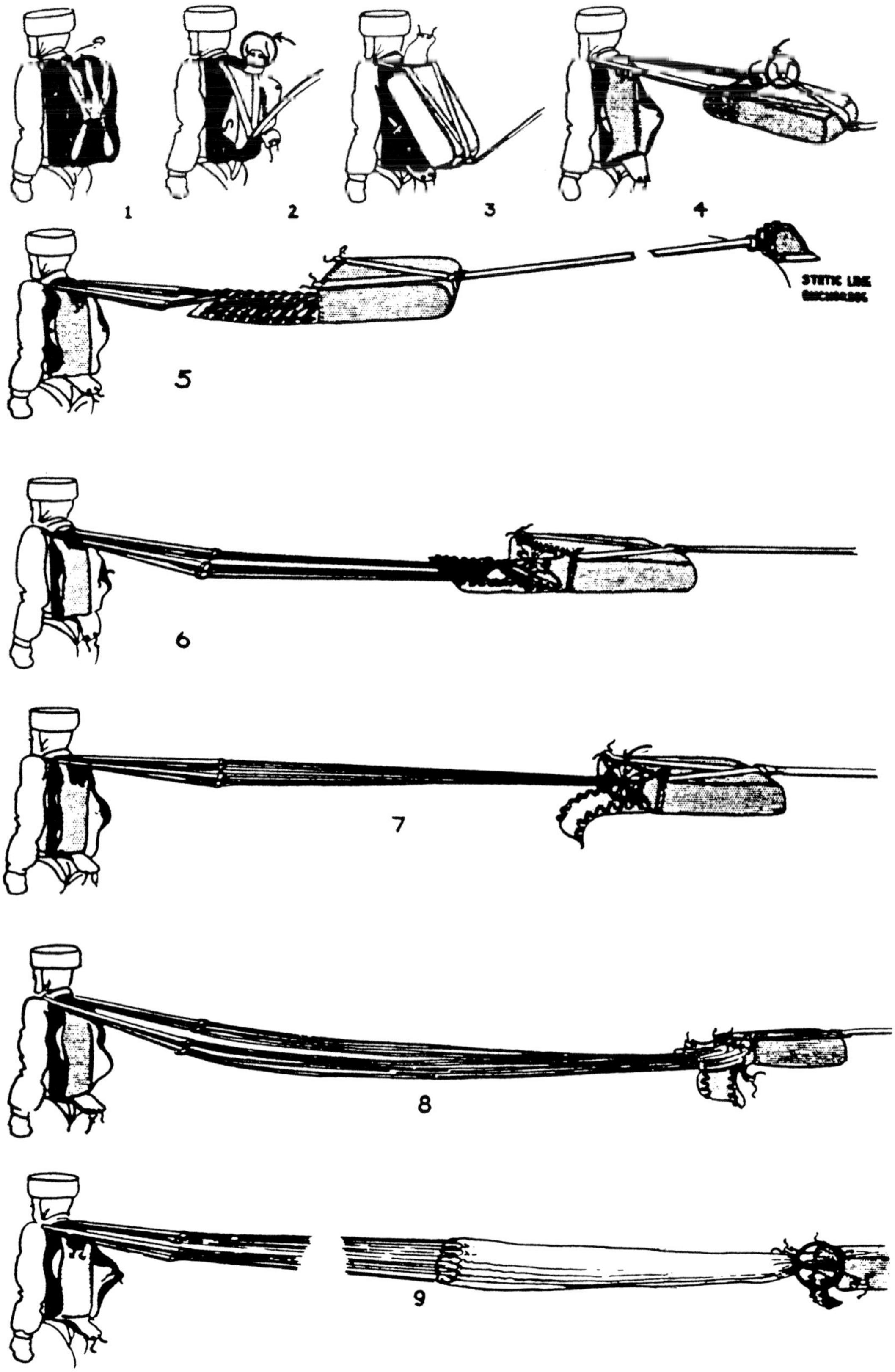

FIG. 53 DEPLOYMENT OF THE "X" TYPE PARACHUTE

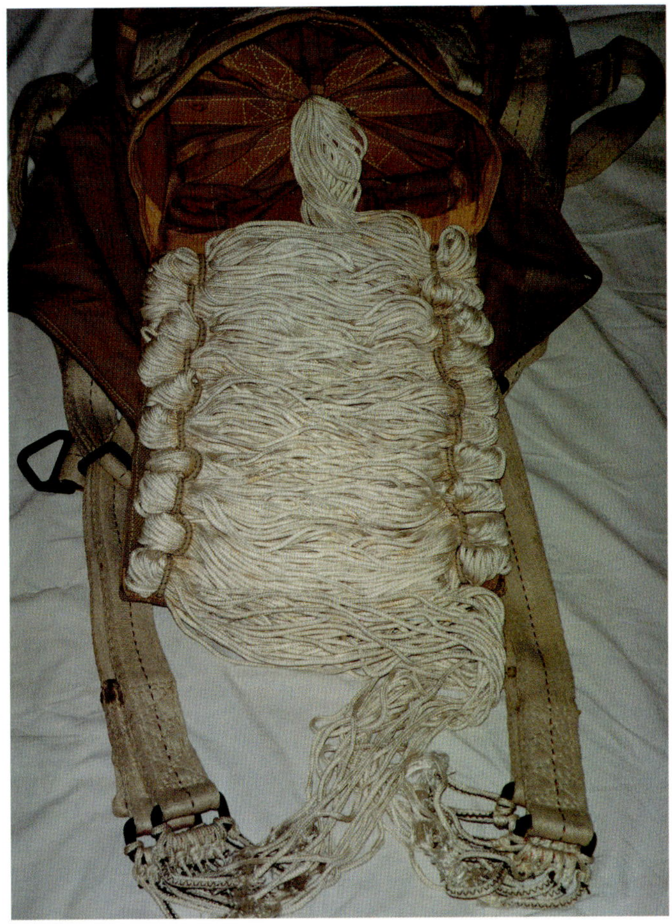

Right: Mk.I inner bag (D-Bag) – early type with bungee cords for S-Line. Note how the canopy is retained by flaps.

Below: Close-up of D-Bag interior.

X Type Mark II outer bag: static line stowed in bottom of container and paid out through slot; suspension lines stowed on top of bag. The X Type can be readily detached from its harness by unsnapping the pack tray/harness loops.

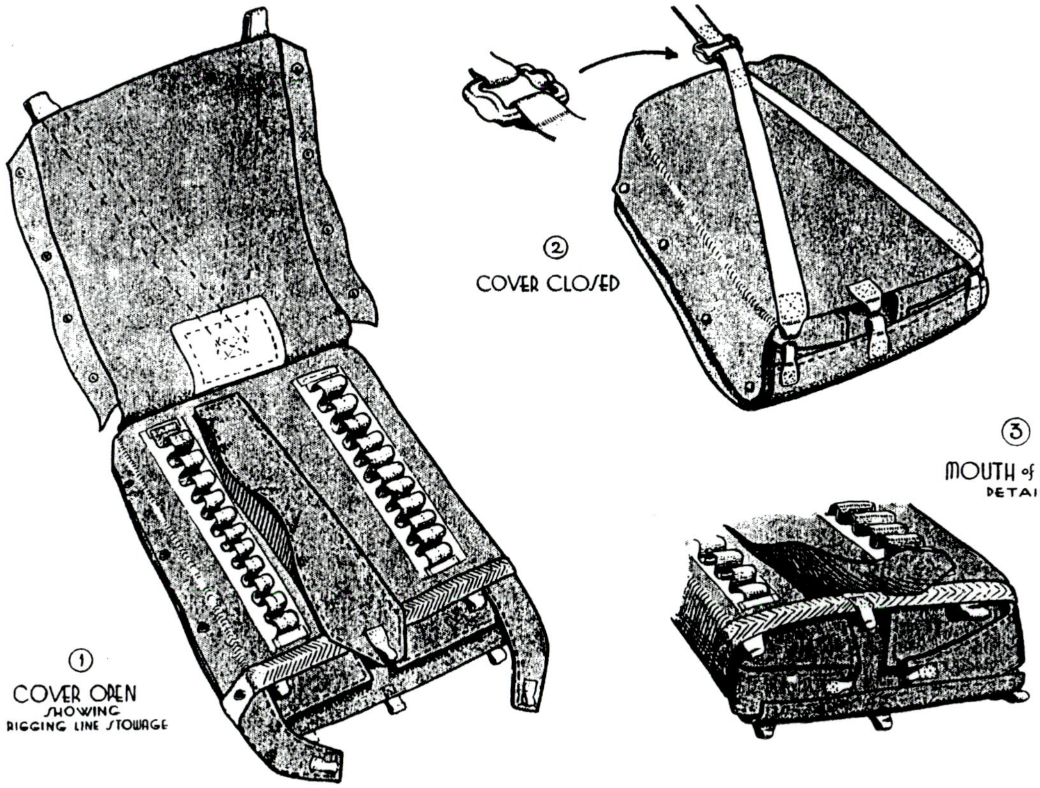

FIG. 51 MODIFIED X TYPE PARACHUTE INNER BAG

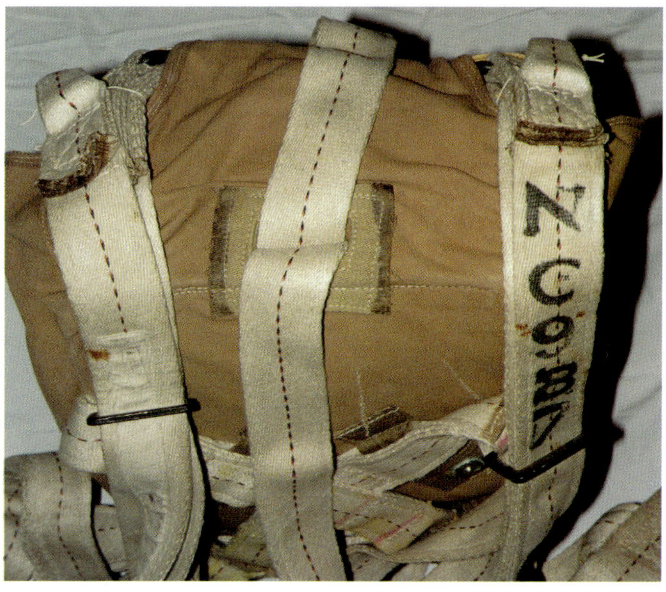

Slot from which static line is withdrawn.

Left: Mk.II outer bag (container) with grommets – much of the D-Bag is exposed.

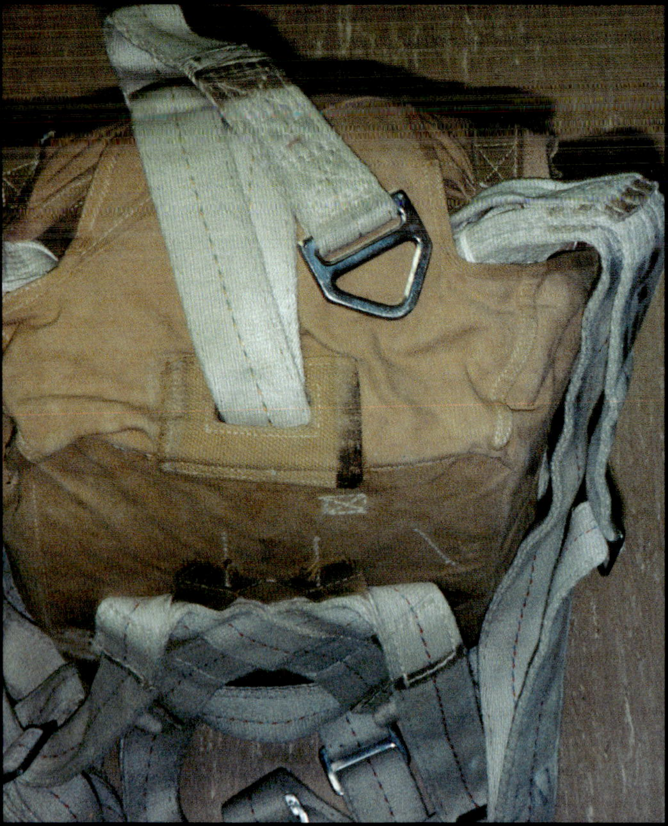

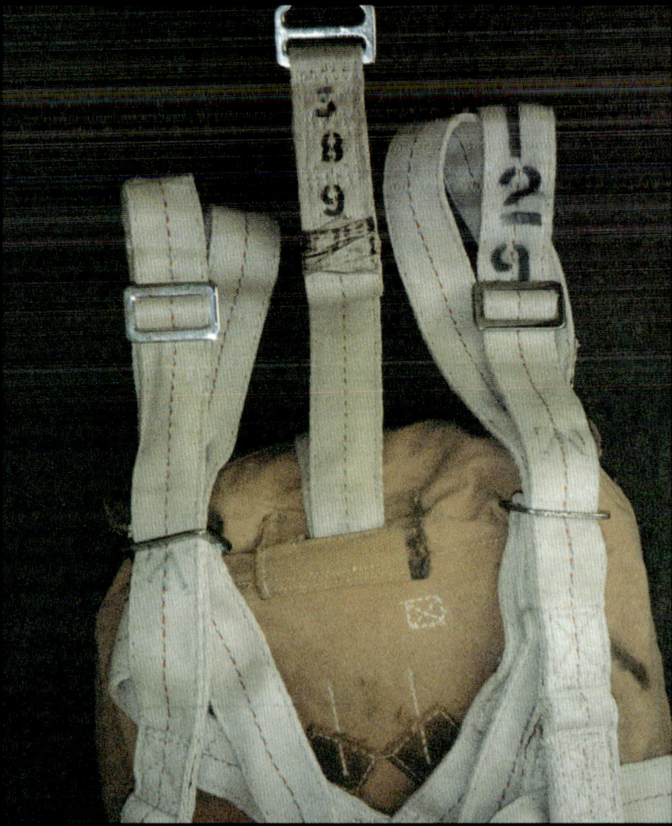

Additional views of X Type Mk.II outer bag showing static extraction slot.

Left: Interior of container showing static line stowage. Right: Close up of static line stowage. Note two Beckets beneath slot should have a break tie.

The Quick Release releases all four lugs and is retained by its own strap.

This Quick Release is captured by its left upper harness lug.

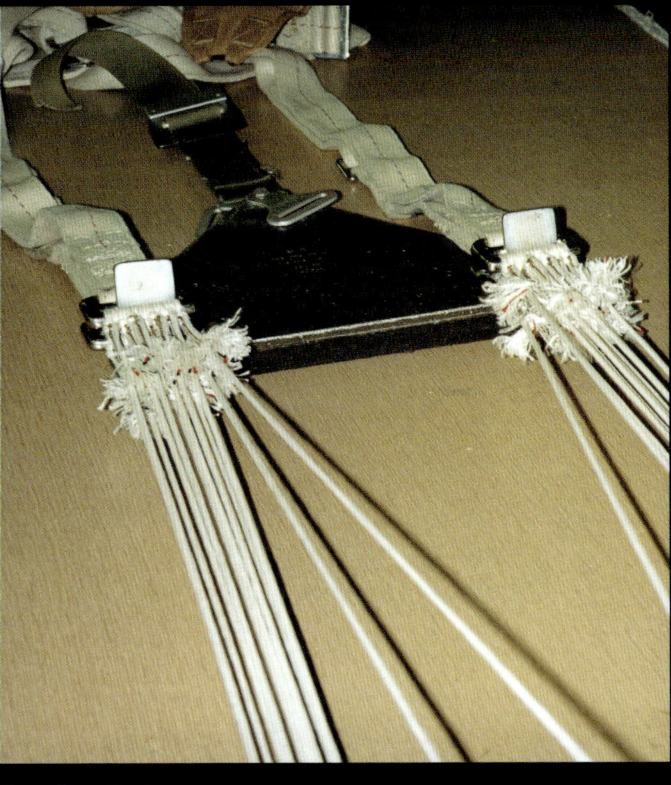

Four line check ensures proper sequence

Wind channel of pleated canopy revealed.

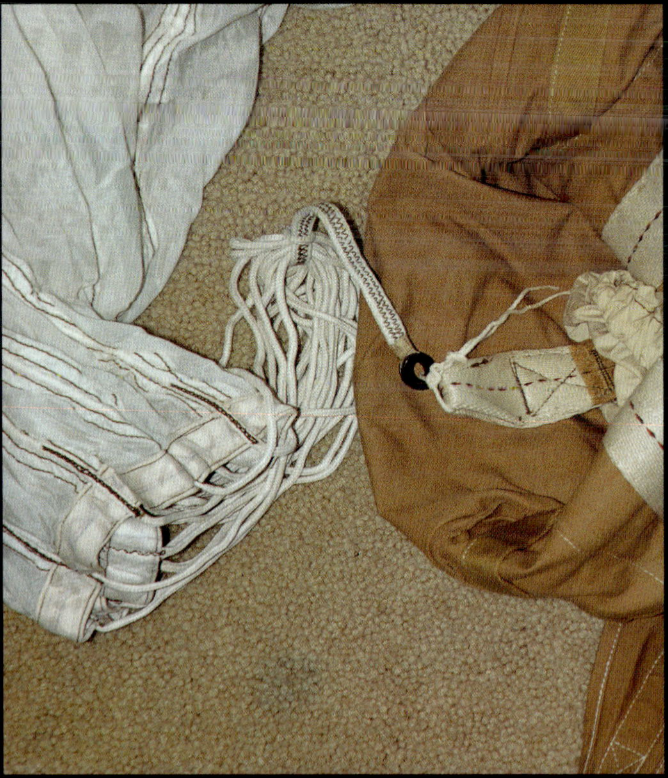

Canopy apex break cord tie.

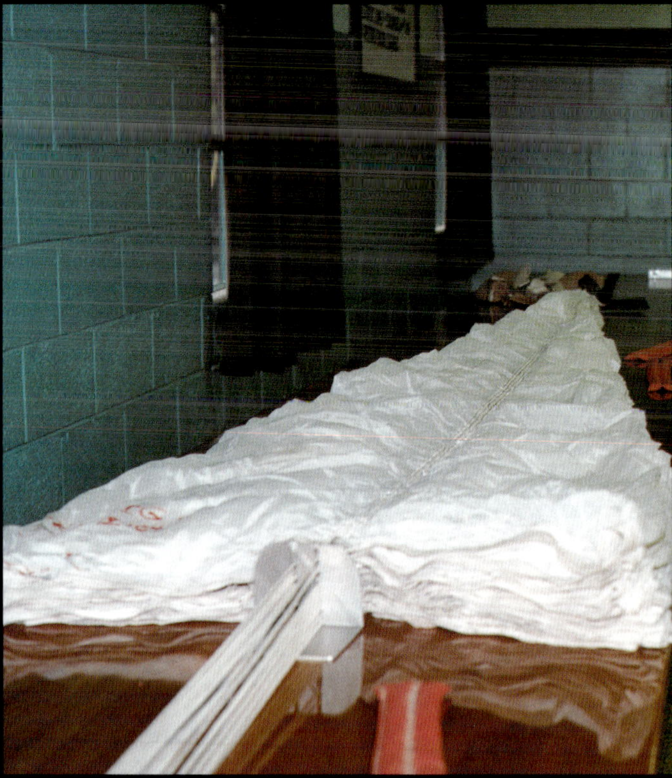

Canopy pleated.

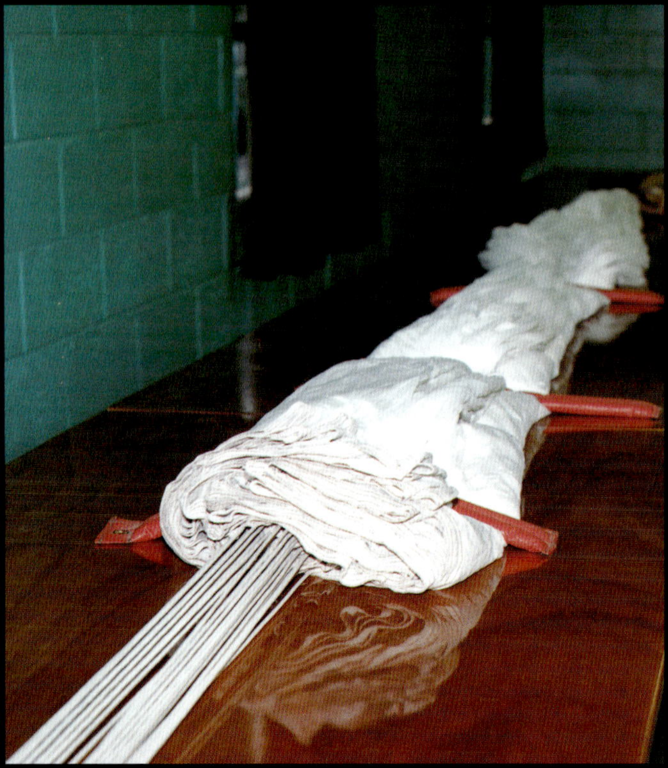

Canopy folded in 1/3 segments.

Apex stuffed into protective sock.

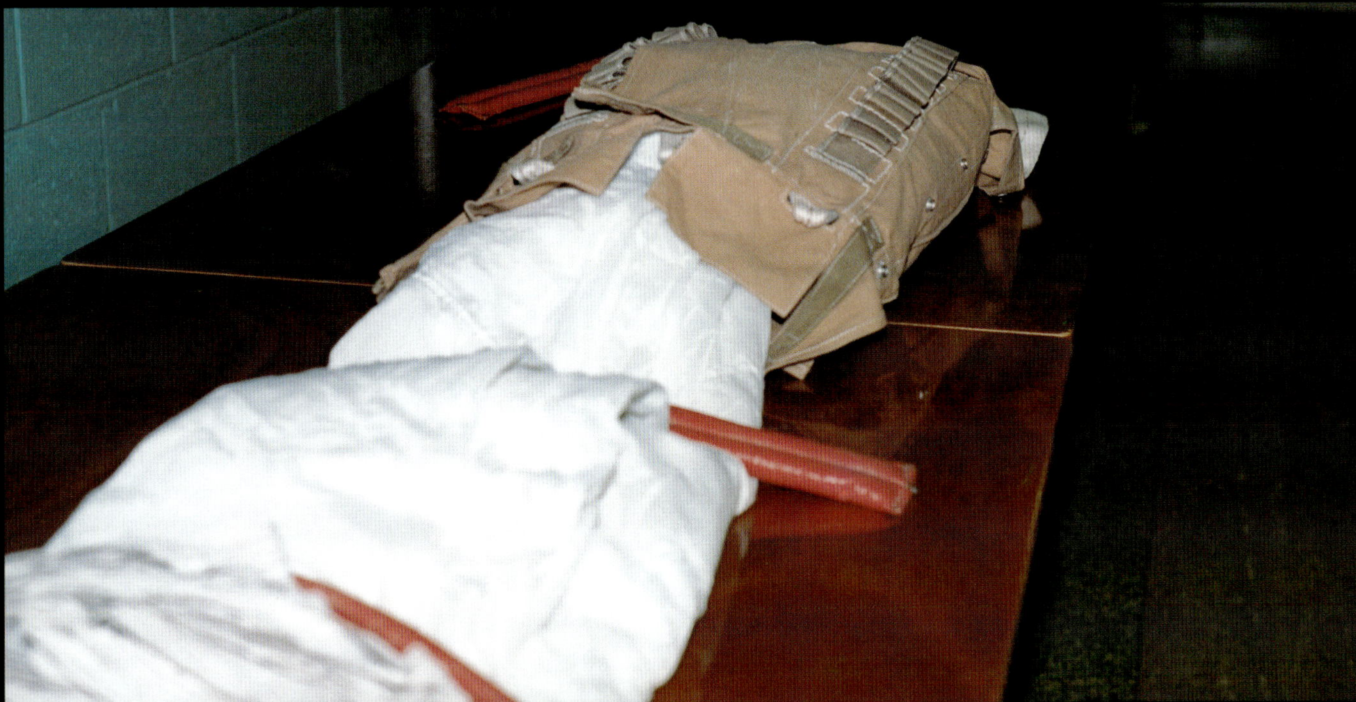

Canopy is S-Folded into bag.

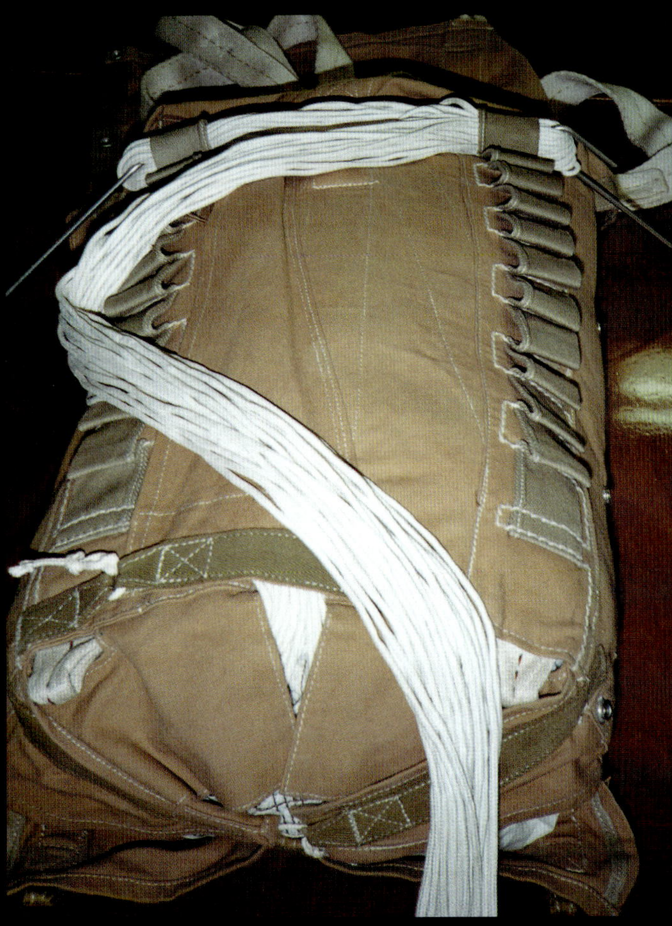

Stowing lines seen from bottom of bag.

First stow with covering flap pulled back.

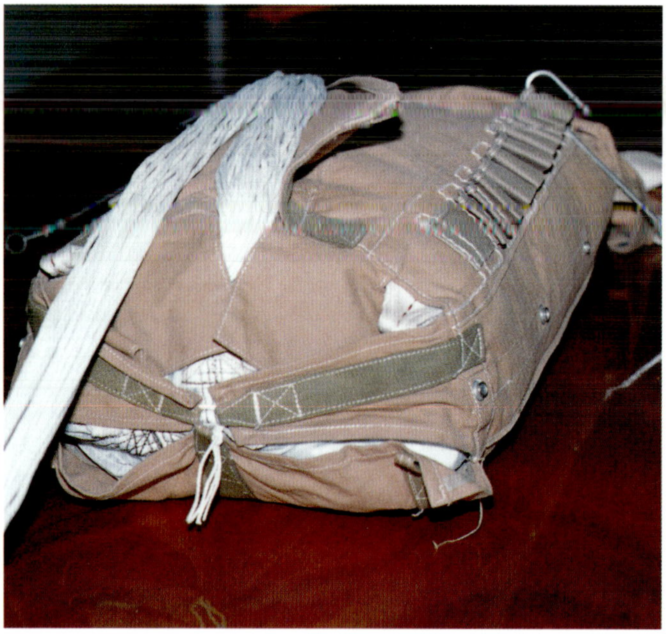

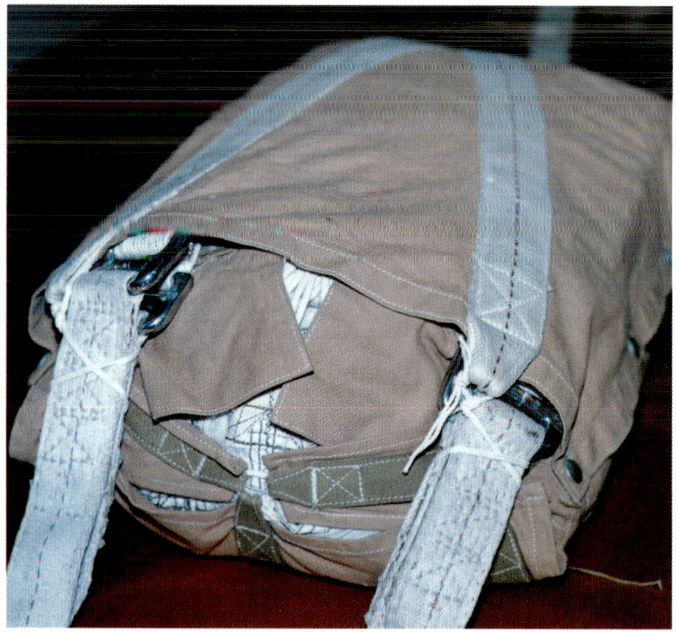

Additional an non-regulation x-ties secure these usually large connector links to the D-Bag.

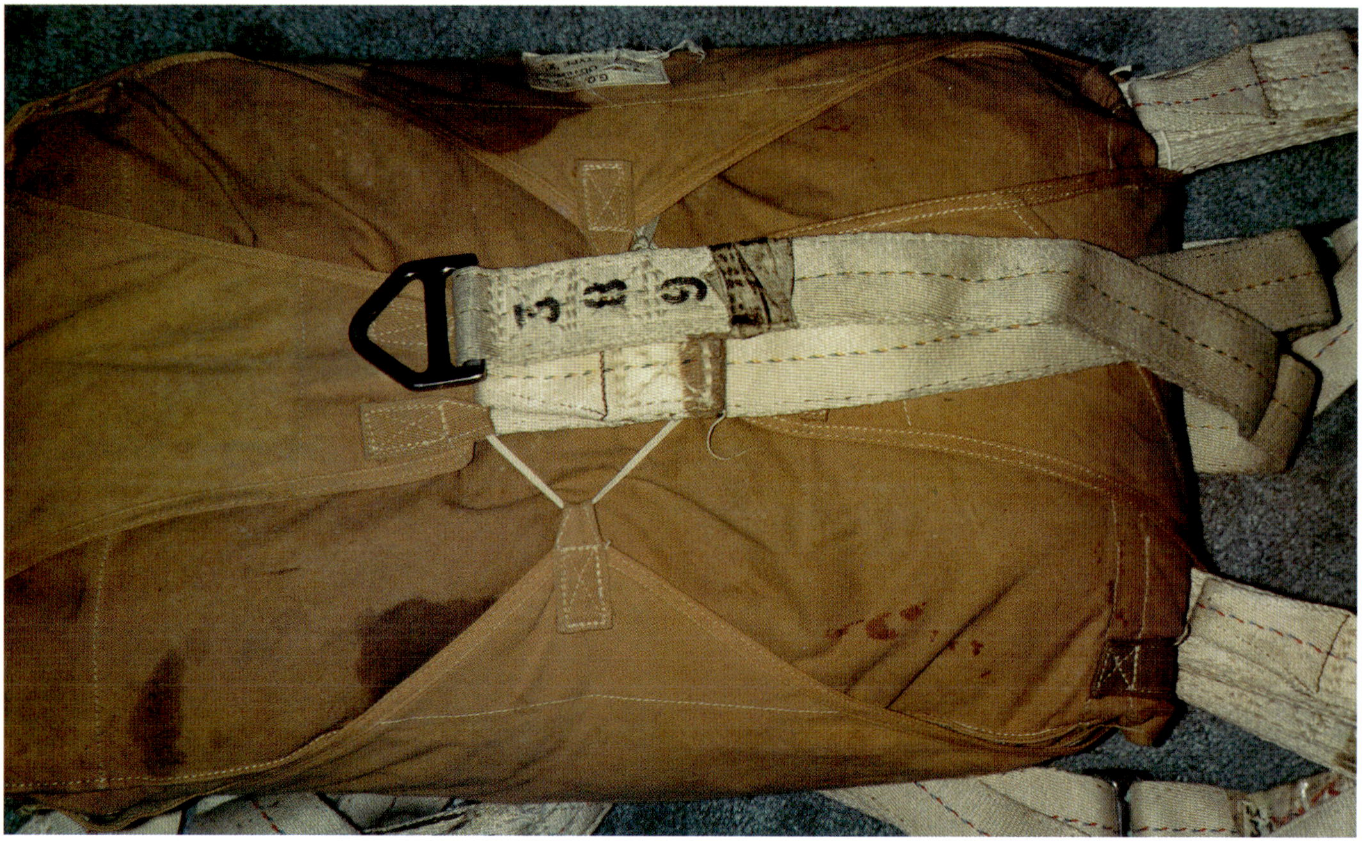

This Mk.II has Becketts for container closure.

FIG. 15 PHASES OF THE FORWARD LANDING (INCORRECT)

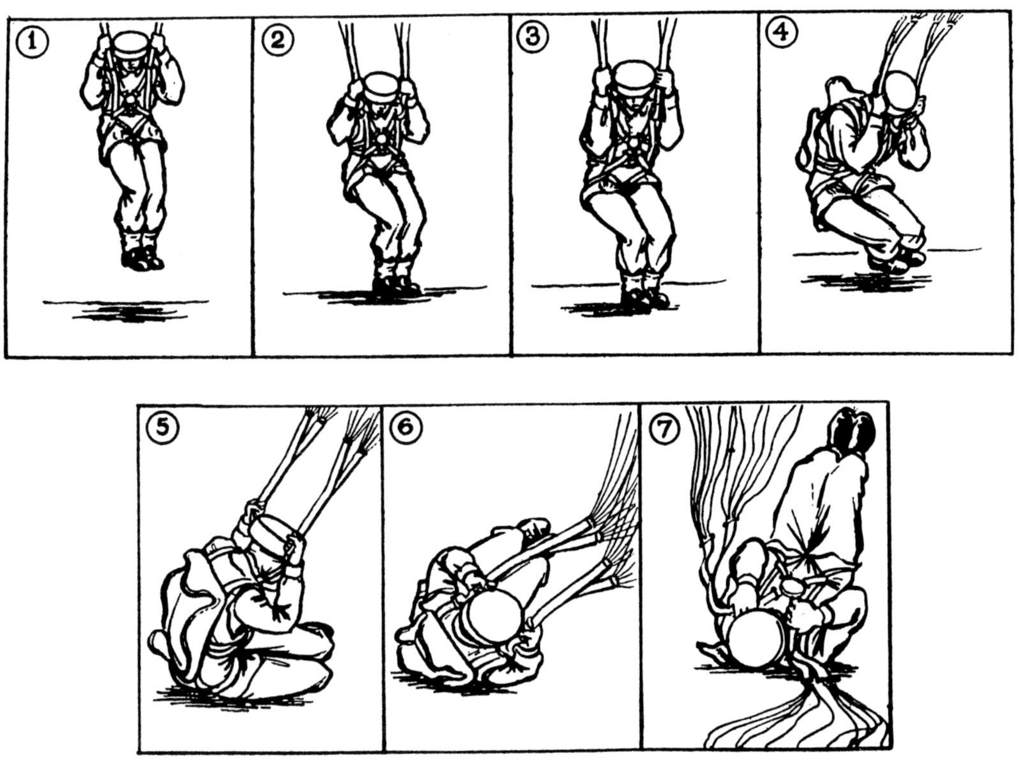

FIG. 16 PHASES OF THE FORWARD LANDING (CORRECT)

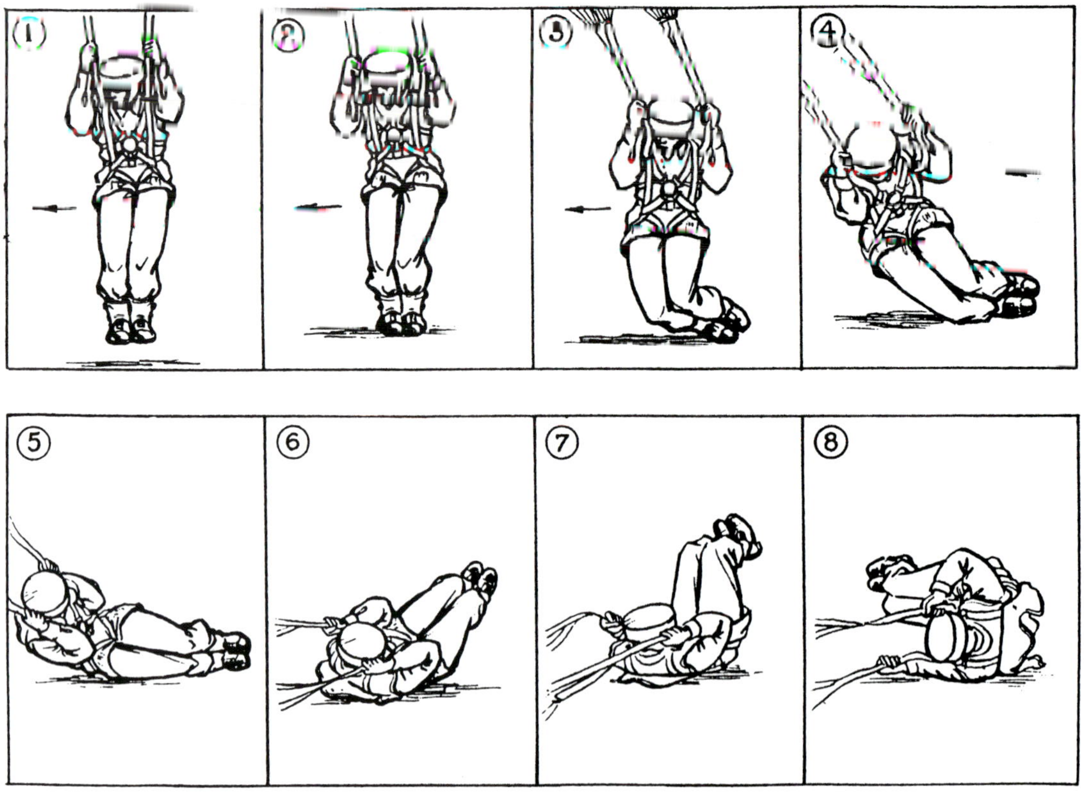

FIG. 17 PHASES OF THE SIDEWAYS LANDING

FIG. 18 PHASES OF THE BACKWARD LANDING

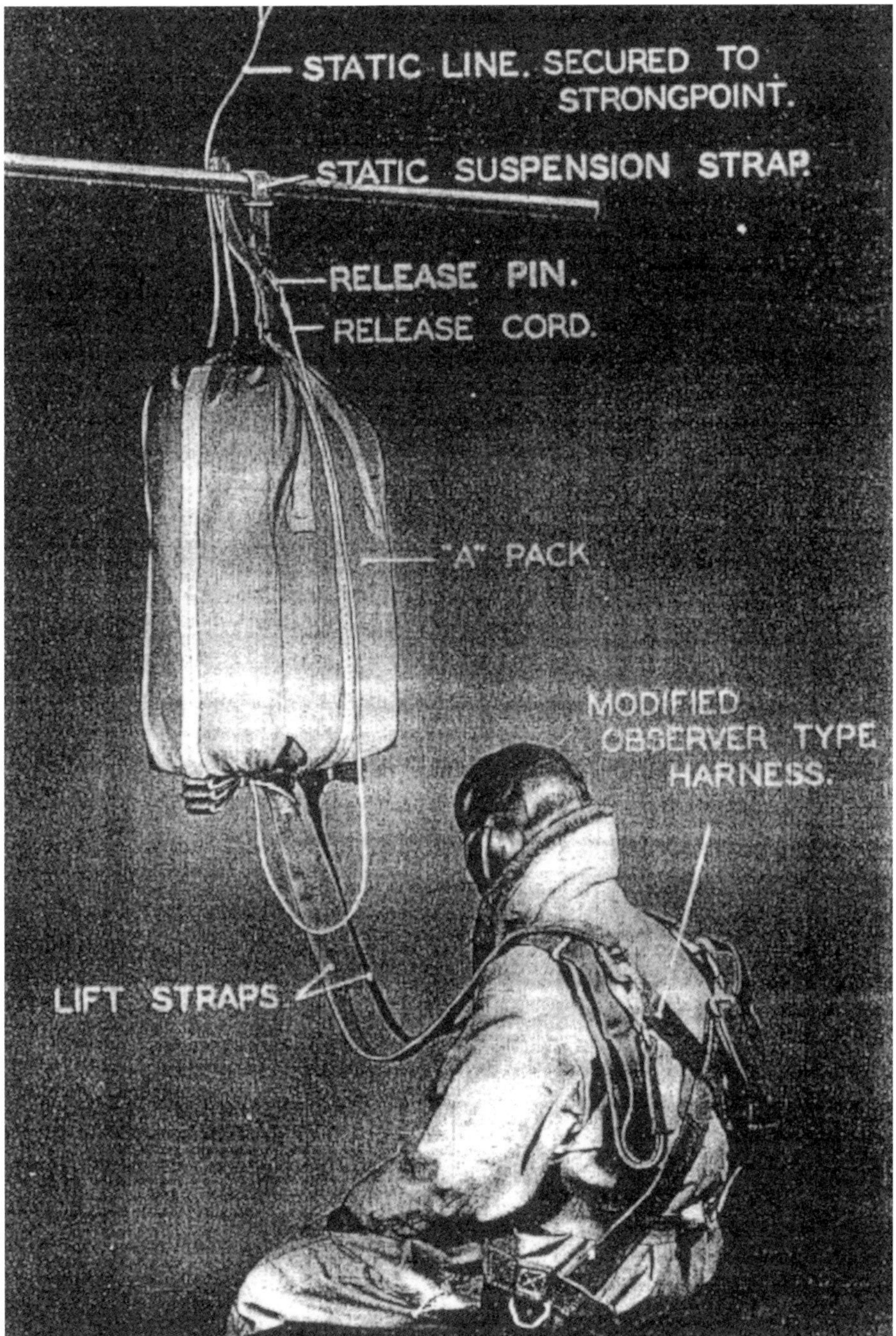

ARRANGEMENT OF "A" PACK AND PARACHUTIST. FIG.

CHAPTER FIVE

United States

Billy Mitchell should be regarded as the spiritual father of the U.S. Airborne Forces. He proposed, in November 1918, dropping the First Infantry Division behind German lines for the anticipated Spring 1919 offensive. It is doubtful that it could have been done given the time constraints.

There was though, a committee pondering the parachute question for aircrew escape purposes. This was the Parachute Development Board chaired by Major E.L. Hoffman and including Guy Ball, A.L. Stevens, J.J. Higgins, Glenn Martin, J.M. Russell, Floyd Smith and Leslie Irvin. The Smith back type was worn on test flights in December 1918. History records that Leslie Irvin made his landmark jump with the device on 28 April 1919.[52] However, it was not until the mid 1920s that serious experimentation in the concept was conducted.

Master Sergeant Edwin Nichols should be recognized as the first U.S. paratrooper. He suggested parachute troops when he was on Mitchell's staff in France in late 1918 and began to work on the project when he returned from overseas in 1919. He was instrumental in developing a back type, static line activation and developed padded containers for dropping cargo. The first public demonstration jump was 29 April 1928 at Brooks Field, San Antonio Texas with six soldiers and a machine gun from seven aircraft.[53]

Claire Chennault, later Major General, was the chief pilot at the time. The Russian military mission led by a Gen. Baranoff was impressed by the spectacle and offered Lt. Chennault a five year contract of $1000 a month, flight status, expense account and Colonel's rank. He turned them down.[54]

The first mass tactical drop was made on September 28, 1929 with eighteen jumpers from nine DeHavillands and three Douglas biplanes. The jumpers dove over the side at 2000 feet and the containers were released at 3000. Four minutes later the troopers had the machine guns firing. The year previous, a demonstration was arranged for the Chief of Staff, MG Charles P. Summerall. He was not favorably impressed with the spectacle, stormed off and heard to be muttering, "some more of this damned aviation nonsense!"

This was only two years after Billy Mitchell had resigned following his censuring court-martial of 1925. General Summerall testified against Mitchell during the trial, so he was neither objective nor imaginative.

In fairness, it must be said that Mitchell lacked diplomatic skills and advanced his air arm crusade by stumping around the country like a politician canvassing for votes and thereby grievously violating military decorum. Worse he proved his point in June 1921 by actually sinking a seized Imperial German armored cruiser, the Frankfort, with Martin bombers in an expeditious manner (twenty-five minutes) and just as he had predicted. He proved that was no fluke in July 1921 by sinking the dreadnought Ostfriesland in twenty-one minutes with six Martins.[55] The man was a threat to the national prestige of capital ships and an embarrassment to the Army with their quaint horse cavalry, coastal artillery and miniscule budget.

Ultimately he was right. Unfortunately, Nichols died in 1935, and Mitchell in 1936, with neither seeing the vindication of their work.

The Mitchell affair had fairly polarized the Army and Navy into pro and anti-aviation camps and the anti-partisans were in positions of authority. In 1929 the War Department issued a directive to Brooks Field to, "stop that nonsense before somebody is hurt." So the U.S. would not be the pioneer in Airborne Forces development.

Brooks Army Airfield, Texas, late 1920s.

Riggers practice mass drop technique over Brooks Field, Texas in the late 1920s.

Parachute research during the 1920s concentrated on aircrew escape methods. Riggers of the Army and Navy were the corps of professional talent in this endeavor. Air Service Order #44 of September 26, 1919 required auxiliary parachutes for any intentional jumps made on Army installations.[56] The civilian world followed this lead when the Department of Commerce decreed in 1930 that all aeronauts making an intentional jump would henceforth perform them with a, "single harness dual parachute pack." The industry had been offering rigs with integral reserves at least since 1924.

This history determined that U.S. paratroops would be equipped with reserve parachutes. They were not alone in that regard. The pre-war French, Poles 1936-1939, Russians and the Japanese Army all had reserves.

Parachutes were classified by the manner in which they were worn and so designated by a prefix: A, S, B, T. Type A was a chest mounted variety and part of the QAC system. QAC was Quick Attachable Chest where the aviator wore a harness that had integral D rings or snaps and when the situation demanded, he clipped on the appropriate chest pack and bailed out. S meant seat type (S 4, S5) closely associated with fighter planes. B stood for back type (B7, B8). These were often provided to aircrews who stood most of the flight, like gunners. T indicated training and meant that it also had a reserve and was intended for premeditated jumps as opposed to emergency escape. Army and Navy riggers had been required for years to make periodic jumps to demonstrate that they stood behind their work and to attempt to sell aviators on the reliability of their escape equipment.

It was proposed back then, but not required, that pilots make some premeditated jumps to prepare them for any eventuality. This was received with somewhat muted enthusiasm as airmen of the day regarded the parachute as an alternate means of dying and parachutists as firmly within the lunatic fringe. (Sadly, this attitude persists today).

The U.S. Army's first troop parachute was actually the T3 system. T stood for "training" at that time as intentional jumps were said to be for training. The T3 was manually activated via a ripcord releasing a twenty-eight foot silk main canopy. The twenty-two foot diameter reserve was permanently attached with short risers merging into the main webs of the harness through steel adapters at the shoulder points. The T3 was demonstrated to, but not used by the Test Platoon. The T3 was converted into the static line operated T4 by changing pack trays. The T4 pack tray was rather shallow and the pack tray cover mostly encapsulated the canopy. The harness hardware was the male-female "bayonet" type (fastener, parachute harness 40A6465, 40A6469) with two snap hooks that secured the reserve pack container. The reserve ripcord is trapezoidal with the locking pins wire wrapped and soldered. Sometime during the war the "T" was understood to mean Troop type.

Navy parachutists, riggers demonstrate pull-off technique with and without reserves during the 1920s.

The T4 static line was originally fifteen feet long. This was attached to the pack tray cover at its center and was temporarily stowed on the back in S-folds with rubber bands. The static line terminated in a simple snap hook (41B21538).

It was proven by motion picture photography that the snap hooks rebounding off each other at the end of the anchor cable could unhook and make the exiting trooper a free faller.[57] Indeed this happened to one trooper who was accused of being so excited that he forgot to hook up. His protestations of innocence were vindicated by the movie camera investigation. The solution to the problem was to simply add a screw in blocking safety pin (42A3519). The pin was tied to the snap hook by a length of string.

The T4 was the system used by the Test Platoon and the 501st Parachute Battalion until the T5 was adopted in the summer of 1941. There was a transition model, certainly experimental, that was a T4 with a T5 pack tray.

The growth of the U.S. Airborne was hampered by equipment shortages. Irving Air Chute contracted to deliver 200 T5 systems a week by September 1941 and 100 each week thereafter. Switlik Parachute Co. promised 3750 by July 1941. Neither company could deliver that many systems, at least initially.[58]

Early in 1940, Herbert Mayer of Reliance Manufacturing offered his company's services to Materiel Command HQ at Wright Field, Dayton Ohio. Reliance was the largest garment maker in the country and naturally suited to parachute production. Samples were submitted and approved in October 1940. Reliance became the largest parachute maker in the U.S. during World War II with two factories in Indiana and one in Mississippi.[59]

The T4 was objectionable because it was awkward to don and doff and even more so when the reserve required repacking, which then was every thirty days. The T4 was re-engineered at Wright Field to become the T5.

Early experimental jump training.

Early parachute and harness.

The classic troop parachute of the U.S. airborne during the war was the T5. It was approved as the standard troop type on 13 June 1941.[60] It is identified by its three separate snap hook harness closures, separable reserve pack, and pack tray flaps that fully enclose the main canopy.

The harness was fabricated from multiple layers of Type VIII cotton webbing which is 1 3/4" wide, has two black center nylon threads and a rated tensile strength of 2900 pounds. Initially the harness was made from undyed, white, web. This changed to a more tactical green by 1944. Almost an entire 100' roll of webbing was required to make a harness. The early T5s had V rings for reserve attachment. These were replaced by 3000 lb. test D rings in order to carry cargo.

The T5 pack tray is rigid being wire reinforced and made of heavy cotton duck fabric which may be dyed khaki, mustard yellow, olive drab, olive green, dark brown or any combination thereof. The belly band, coming around from the jumpers right side, and connecting into a adapter on the left, was routed through web loops on the reserve back and served to hold the reserve pack in place if it was activated. The belly band also could help secure a weapon to the jumpers side. These were sometimes cut from the tray at the jumpers discretion in order to speed doffing the rig after a combat drop. Reserves were also optional in some units during a big operation. The static line is thirteen feet long plus/minus four inches. Like the T4 this started out as a fifteen foot length of webbing, but tests involving motion picture camera work showed conflict with the battery box handle on the tail cone of the C-47. So the static line was shortened a foot at a time until it could not connect with that protrusion. It also prevented the canopy from striking the horizontal stabilizer.[61]

The pack tray cover to which the static line is permanently affixed provided a strong loop for connecting the apex of the canopy by break cord. This break cord could be any suitable item as long as its cumulative breaking strain was 100 pounds. That is the apex break tie could be one, two or three turns of something that took 100 pounds to fail such as one loop of No.100 silk tape or three turns of 16-5 linen cord.[62] This apex tie was crucial to prompt deployment as it was intended to draw the canopy out before separating.

The T4, T5 and T7 all developed in the same manner. They opened canopy first. The canopy being dragged directly out of the pack tray by the static line and into the slipstream, before the suspension lines were stretched out. Even then, it was recognized that having the lines deploy before the canopy was better, safer and easier on the jumper.[63] There was no perceived time to devote to experimenting with a new system. The T5 was regarded at the time as a stop-gap measure. This system of canopy extraction via static line secured with break cord on a laced cover dates back to Charles Broadwick's 1913 coat pack[64] and was patented by Leslie Irvin on December 16, 1918.[65]

The tray cover, also called the "pull out panel", was laced to the pack tray with one or two turns of any cotton cord that would have a breaking strain of fifty pounds.[66] Two other ties served to hold the risers in place within the pack tray. These were across two grommeted tabs. Finally, the risers were hand tacked to the bottom of the tray with thread. The suspension lines were S-folded and stowed on the bottom of the pack tray with rubber bands.

No troop parachute was faster opening than the T5 and none was more punishing to the jumper. The canopy was ripped from its lair by the static line and flung into the prop wash before the suspension lines could be extended. This abrupt deceleration always left bruises on the trooper's chest and back where the bulk of the shock is taken. The harness works like Chinese handcuffs compressing the load laterally.

Greatly simplified, Floyd Smith's harness is a web sling that a person sits in. The Main Lift Webs run vertically down the front torso and make a continuous loop around the buttocks.

Early mass drop. Dated April 11, 1941, these 501st Bn. troopers are using the T4 system.

The back portion is closed by an X-shape of harness webbing and the front is secured by adjustable leg straps and a chest strap. The whole affair suspends the individual rather sensibly from his shoulders. The T5 opening shock was such that if the plane exceeded 150 mph, it could knock troops out. Canopy damage and malfunctions multiplied directly proportional to any increase in airspeed.[67]

The Test Platoon members learned early on that best exit form included ducking ones head until the chin rested on the chest to avoid being crowned by connector links. Risers swinging up could and often did pry helmets off. Practice Drop Zones became littered with rifles, helmets, canteens, packs and anything else that was not securely tied to the jumper. Trucks were filled with this debris following the obligatory police call after a mass tactical drop.

Since the T5 impressed such tell-tale stigmata on each user, the OSS sent their field personnel to the British school so that when they were launched into occupied Europe by the "X"

Demonstration drop at Ft. Belvoir, VA, October 3, 1941.

Type, they would not be betrayed by the bruises. OSS troops so trained were accorded the honor of wearing British and "Special Forces" wings on their dress uniforms if they belonged to Special Forces HQ.[68]

The T5 reserve had a twenty-two foot diameter canopy until mid 1942 when the T5A with twenty-four foot diameter replaced it. The early T5 reserve is quickly identified by flap extensions of the container that served for carrying. These flaps had a vertical web strap on the back through which the belly band passed. Later versions deleted these flaps, the web straps were moved to the center and a carry loop was added on the right side.

In mid 1943, carry loops were sewn to both sides and the top. These additions were doubtlessly required when too many people reached to pick up a reserve pack and grabbed the wrong handle.

The T5A was equipped with a spider type pilot parachute which has continued in use to the late 1990s.

Both the canopies were made from nylon twill fabric. The war with Japan was a boon in this regard as nylon is a far superior parachute material than silk. Nylon was proposed as a silk substitute as far back as 1939 and the war accelerated its adoption.

The T5 main canopy was a nominal twenty-eight foot flat circular. Canopies laid out flat will usually measure little over twenty-five feet, but aloft will spread to about twenty-eight due to stretch factor. Wartime procurement for main and reserve canopies was 50/50 percent white and camouflage.[69] The reserve mix continued until 1953 when white was deemed standard in order to spot reserve canopies, and presumably malfunctions, faster.

There was a shortage of T5 mains in North Africa before the invasion of Sicily in July 1943. The Army had been operating a jump school for replacements and training a French battalion. The terrain, climate and particularly the cactus all combined to ruin parachute canopies and especially pack trays. Some 200 Navy Dual Purpose systems were rushed there where they were affixed to T5 harnesses. The ripcord cable and housing was left on the rig disengaged and a jury-rigged static line was stowed vertically on the outside of the container.[70]

The experience at Normandy expedited a major modification to the T5.[71] A certain scandal-seeking journalist (Drew Pearson's syndicated column 'Washington Merry Go Round', 23 June 1944) learned that eight paratroopers drowned in a lake at Camp Mackall early in 1944[72] and deduced that the Army leadership had neglected to provide quick egress parachute harnesses. Even *Life* magazine had a pictorial on the question (June 19, 1944).

This was not true as the T7 with a single point release box had already been adopted, but was not yet available in quantity sufficient for all the Theaters of Operation. Priority on the quick release went to aircrew escape types, and rightly so. Never mind, sensationalism and the fevered emotions of wartime prevailed. Sensitive to accusations, SHAEF commanded that all T5 harnesses would be retro fitted with quick release assemblies purchased from the Irvin UK firm.

There were two versions of this. The first was done on a crash basis by removing the chest strap and installing the British produced release box that discharged all four lugs. The leg straps were left intact and extensions were provided with either a snap or V-ring on one end and a QR box lug on the other. These snapped into the appropriate harness piece were routed around the main lift web and into the QR box.

T5s thus converted were referred to as the YT7. These rigs were only available to units of the 101st Airborne Division for the "Market-Garden" drop into Holland. The 82nd made do with standard T5.

The second modification was made in the winter of 1945 and was far more substantial. This involved sewing new extended leg straps into the saddle of the T5 harness and turning it into the T7. These can be identified by the white webbing leg straps on otherwise green harnesses seen in pictures of the 17th Airborne jumping from C-46s in Operation "Varsity". They had 10,000 of these YT7s for the jump over the Rhine.

The T7 proper did appear in the ETO in early 1945 and these equipped the 82nd Airborne then at Mourmelon France awaiting the call to drop on Berlin. They made practice jumps with 2,280 T7s while the 17th Airborne Division made do with the YT7.

A rare variant of the T5 was produced of undyed fabric – that is solid white – for test unit purposes on specifications from Wright Field. The intent was to provide a stark contrast for photography. Some of these were made available to the experimental arctic airborne unit (Co.B 503rd P.I.Bn) that trained in Utah from January to March 1942. Perhaps as many as 100 were also sent to the First Special Service Force when a company of that unit was placed on standby to make a drop onto a beach on Kiska in the Aleutians.[73]

The incorporation of the quick release box into the U.S. harness has an interesting history of starts and stops. Capt. William P. Yarborough was an early promoter of the device and arranged the purchase of 300 units for experimentation.

In July 1941 the T5 with Quick Release box was to be demonstrated in front of the Chief of Infantry branch MG Omar Bradley by dropping a test dummy. At altitude, Capt. Yarborough and Lt. Ryder manhandled the heavy inert load to the door, rolling it over on its belly. They were unaware that in so doing the release button had been rotated and compressed.

The dummy was pushed out, the parachute developed, but "Oscar" slipped his bonds free falling to a dusty impact seconds later. The observing dignitaries were then convinced the proposed modification was unsuitable and no remonstration from Capt. Yarborough would dissuade them.[74]

The official report recorded this episode: "The Irving Quick Release device provided a method of immediate release; however tests disclosed that it was deficient in not providing the necessary safety factor and could be inadvertently released by the parachutist, thereby releasing him prematurely while descending."[75]

The quick release issue refused to go away and was revitalized in April 1943 following the first operational drops in North Africa. Col. Edson Raff of the 509th Parachute Infantry Battalion brought back a quick release device from a German 'chute. This German box had a locking fork that prevented the release button from being depressed enough to activate.[76]

Development continued toward what became the T7 and on 10 October 1943 harnesses with QR boxes were delivered for test. The lag between adoption and procurement was attributable to production retooling and competitive drain to aircrew escape types.[77]

The exit technique began with the jumper standing erect in the door in a slight crouch, looking straight ahead with his arms extended so that his palms were against the airframe skin below his waist. At the tap and command "GO!" he would spring up and out simultaneously locking his legs together, tucking his chin onto his chest and placing hands on both ends

Captain James Coutts inspects troopers of Co.B. 501st Parachute Bn., April 11, 1941. These T4s were produced by Irving Air Chute of Buffalo, NY. The static line hooks were temporarily tied to adapters for convenient access.

T4: the reserve has risers and is permanently attached to harness; main tray cover completely encloses canopy; bayonet hardware on harness.

A transition model: T4 with a T5 pack tray.

Right: Major General William Lee wearing the early balloon cloth coverall, Riddell helmet and T4 system. The leg and chest straps are closed by "bayonet" hardware. The 22' reserve container is held fast to the jumper by two snap hooks while its canopy is attached to risers that are permanent extensions of the harness. – an awkward arrangement that was soon dispensed with. Lee is regarded as the "Father" of the U.S. airborne and only the detection of a heart condition prevented him from leading one of the divisions into combat during World War II.

of the reserve with his elbows clamped tightly to his sides. An elbow out will induce a spin in the slipstream, which is a torrent of air. This tucked position clears the back for clean withdrawal of the deploying canopy and reduces the chance of riser burns, fouling lines and being smacked by a connector link.[78]

The jumper started a three second count at his leap by one-thousands, expecting to be interrupted before "three thousand!" by the arresting canopy. The T5 left little doubt before three seconds elapsed.[79]

He would then check his canopy for roundness or damage. If he should reach the three thousand count without any jolt, he was taught to rip open his reserve.[80] Training jumps were made at 1200 feet AGL, so there would be time for a reserve to inflate.

The anchor line cable ran down the center ceiling of the C-47 cabin. Jumpers controlled the static line by grasping the snap hook and shuffled to the door with the left foot forward. The static line had to run freely over the left arm and they were conditioned to not release the snap hook till they were in the door. Misrouted static lines caused towed jumpers and traumatic injuries.

Jump Commands during World War II

1) Stand up
2) Hook up
3) Check equipment
4) Call off for equipment check
5) Stand to the door
6) GO![81]

The above was drilled into the trainees until it was second nature and on operational jumps the troops piled out of the transports as close and as fast as they could in order to (1) make the drop zone (2) get out of that flying FLAK[82] magnet. It was soon found to be impossible for the troops to exit too close together.[83]

The landing technique as taught initially in the school was a forward somersault with the feet spread. It was considered desirable to be facing downwind so that lateral drift would aid in the tumble. This method was what agile jumpers and barnstormers had imparted as the preferred way, but resulted in too many injuries.

Capt. Yarborough and Lt. Ryder wrought a big change in 1942 by modifying the landing posture to both feet together. Major James W. Coutts brought the PLF back from Britain where he had been liaison officer and this was incorporated into the program at Ft. Benning in late 1943.[84]

First Special Service Force troops undergo parachute training equipped with early T5 systems but not with the proper M1C parachutists helmet.

Jump School was a four week course with each week called a stage. "A" Stage was devoted to calisthenics, judo, rope climbing and running. "B" Stage mixed tumbling, mock door exit drill, 3' tower, landing apparatus and packing. "C" Stage was more tumbling, mock door, wind machine for drag recovery, shock harness, the controlled fall tower and the 250' free tower. "D" Stage was packing and jumping. Packing was deleted from the course after a WAC detachment was posted to the school in 1943.

In 1943 main canopies began to be made of speckled pattern camouflage nylon fabric. This pattern is composed of light and dark green irregular dots arrayed on a field of khaki. This same canopy was incorporated into the T7 main (42J2019-2). The nylon canopy has twenty-eight gores of four panels each. The fabric of each panel is placed on the bias, that is angled at forty-five degrees, for strength. This also was believed to limit tears and divert them to a seam. The nylon suspension lines are seven strand core with thirty-two strand braided sheath and rated at 550 lbs. Total length from link to link is seventy-two feet. The canopy apex is 18" reinforced by tubular nylon tape of 3000 lb tensile strength. All of this data is according to the manuals. The canopy is a nominal 28' diameter. The full 28' being attained by stretch factor on full development.

The T5 pack tray cover is functionally identical to he T7 and was interchangeable. It should be noted that the drawing numbers first two digits are the fiscal year of adoption which then began each 1 July. Date of manufacture can and often will differ from the drawing number. There was overlap between

This trainee wears an early T5. The 22' reserve container has no carry loops, and a square ripcord grip and side flaps through which the belly band is routed. The Riddell football helmet gave more substantial protection than the cloth aviator's type. Circa 1941.

45 BASIC FIELD MANUAL

Articles carried in pockets—Continued.

	Pounds
Left waist pocket:	
Grenade, hand, fragmentation	1.25
Right leg pocket:	
Pistol, automatic, caliber .45, loaded with a clip of 7 rounds	2.762
Two extra clips, each loaded with 7 rounds	0.962
Handkerchief	0.053

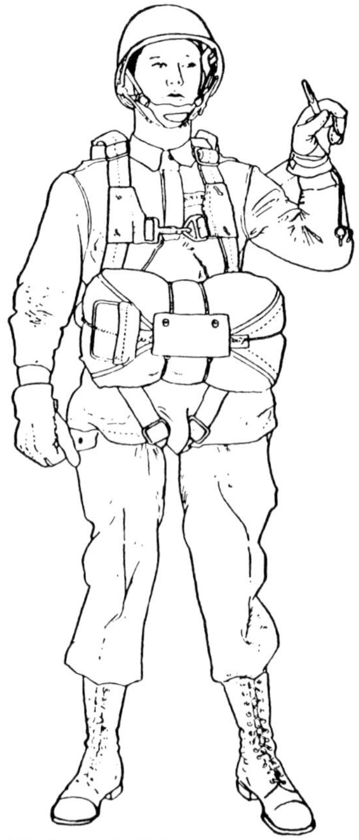

FIGURE 2.—Individual parachutist wearing T–5 troop type parachute and complete jumping equipment.

36

TACTICS AND TECHNIQUE OF AIR-BORNE TROOPS 45

Articles carried in pockets—Continued.

	Pounds
Left leg pocket:	
Rope, parachutist, 20-foot length	1.0
Packet, first-aid	0.234

NOTE.—If the flight is to cross extensive water areas, the troops should be equipped with life preservers and be instructed in their use.

FIGURE 3.—Side view of individual parachutist, showing main canopy, pack on back, and reserve para hute on chest.

37

Illustration from manual FM 31-30. Shown is an early T5 with 22' reserve and 2nd Model static line snap hook, its locking pin tethered to a string.

production of the T5 and the T7. Example: a T7 dated August 1944 and a T5 dated October 1944. There was also a T6, it was experimental only with a novel harness and a flat reserve pack.

Airborne Command recommended the quick release box be standardized and replace the T5 on 26 November 1943.[85] The T7 was standardized 13 March 1944[86] and was in use at Ft. Benning by mid 1944. The T7 was shipped to the ETO in November 1944 with priority going to the 82nd Division.

The T7 can be regarded as an improved T5. The T7 harness is constructed of layers of type X cotton webbing, which is 1 3/4" wide, 0.130 – 0.160" thick, 5000 lb tensile strength and has no color code. The most prominent feature is the single point release box with its round release button. T7s , like T5s, came in a variety of colors and shades. The T7 pack tray grommets were finished in flat black or green. The T7 was slightly modified in 1948 by the addition of puckerbands (*taschengurts*) to the reserve skirt and became the T7A. The T7 soldiered on being gradually replaced by the T10 starting in 1952 and was completely phased out in November 1954.

T5: 22' reserve with sideflaps, no pilot 'chute, square or cloverleaf ripcord grip; white harness made of Type VIII webbing; V or D rings on harness for reserve; 2nd, 3rd or 4th Model static line snap hook; canopy undyed white.

Early T5 reserve container for 22' canopy. Richard Switlik was instrumental in convincing the Army to adopt a separate reserve pack.

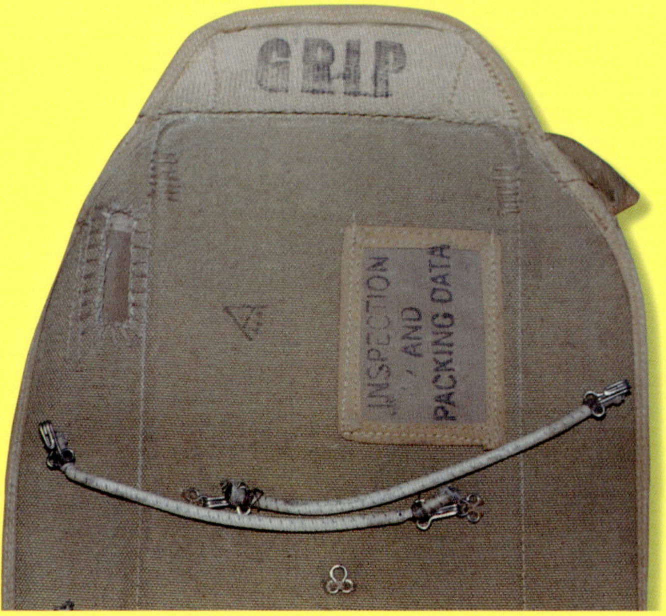

The Third Model static line hook – 42B 3232 – was a short lived transition between the two simple snap hooks that preceded the familiar 43 A9502 sliding sleeve type still in use today. This model can be called the "humpback" as its body is larger than the gate. This hook is massive at 6.5 inches long and a full pound in weight. – in contrast, 43 A9502 is 1/2 lb. The gate slides vertically and is locked by a spring pushed button like its progeny. It is made of two sheet steel plates riveted together with a reinforcing C shaped link sandwiched within.

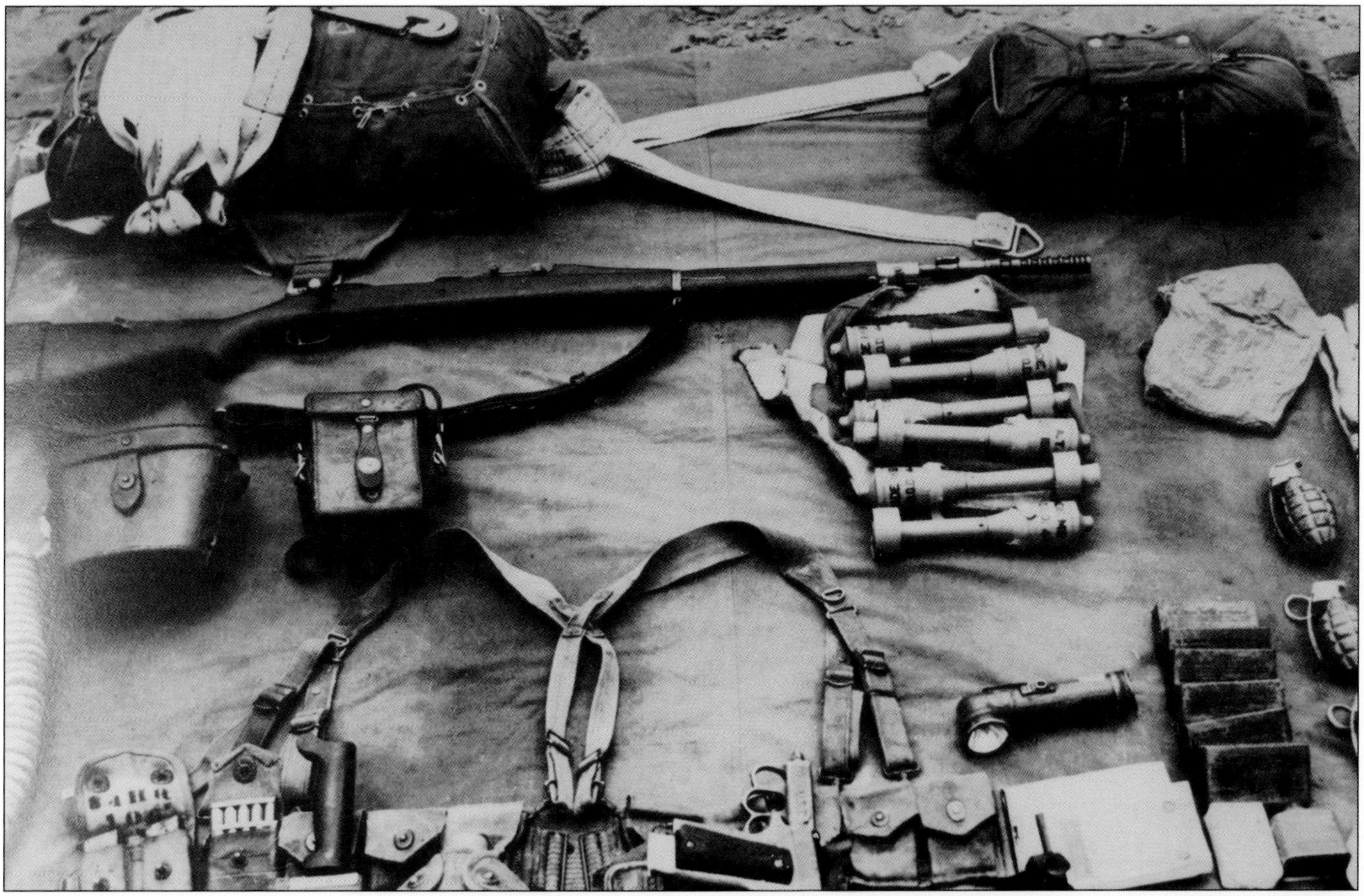

Above: This early war (North Africa) equipment layout shows the T5 to good effect. Note the 3rd model static line snap hook, 22' reserve with cloverleaf ripcord handle and side flaps.

Right: The last static line snap hook was flat sided and only differs from the current model MS 70120 in lacking a safety wire hole under the gate sleeve.

A trooper climbs into a Skytrain in this famous photo associated with the Normandy invasion. The trooper wears a T5A system made of green dyed webbing with the 24' reserve. Note how the belly band has been routed through the right carry loop in addition to the normal reserve retaining straps. The belly band is especially useful here in securing his M1 Thompson submachine gun. The SMG's sling is most likely routed around the reserve D-ring. His magazine pouches are thoroughly taped to secure the heavy magazines. The position of the horizontal back strap over his canteen means he probably got a bruise there on opening shock. The knee areas of his M42 jump suit have been reinforced as are the pockets.

T5A: 24' reserve with pilot 'chute; reserve container has one or three carry loops; harness white or green; reserve ripcord cloverleaf shaped; canopy white or camouflage.

OCT 12 1944
TOP

AIRBORNE

Left: The manufacturer data panel could be a sewn on piece or stamped and on either side. Right: The webbing from which the harness was made – Type VIII – was treated with a mildew resistant chemical that turned yellow with age.

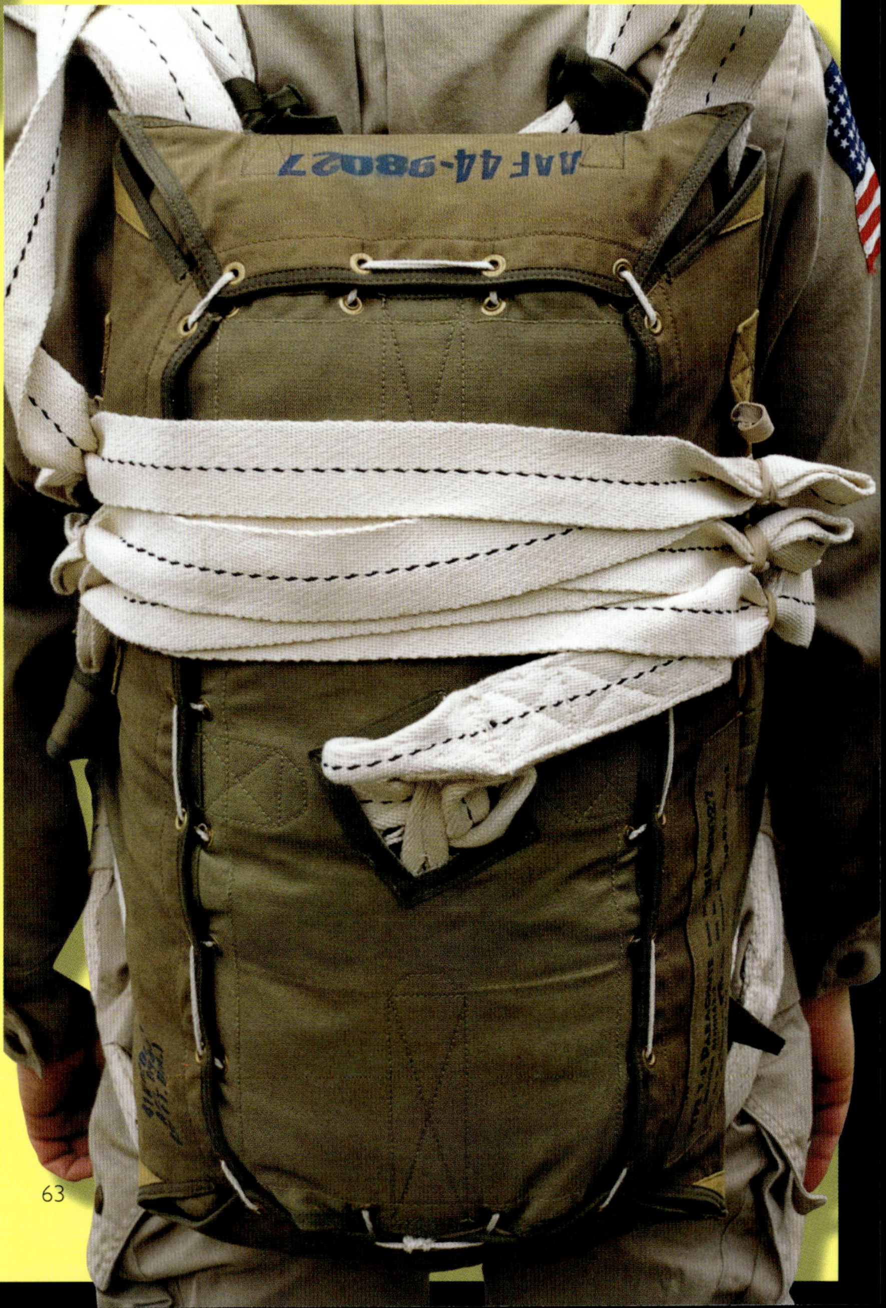

T5A: Note 24' reserve with cloverleaf ripcord grip and 4th Model static line snap hook, and undyed webbing. The Army Air Force was the proponent agancy for all parachute equipment. The pack tray was stamped with these contract numbers in blue ink.

T5A reserve container. Though marked T5, these are for the 24' flat circular canopy. Note three carry grips.

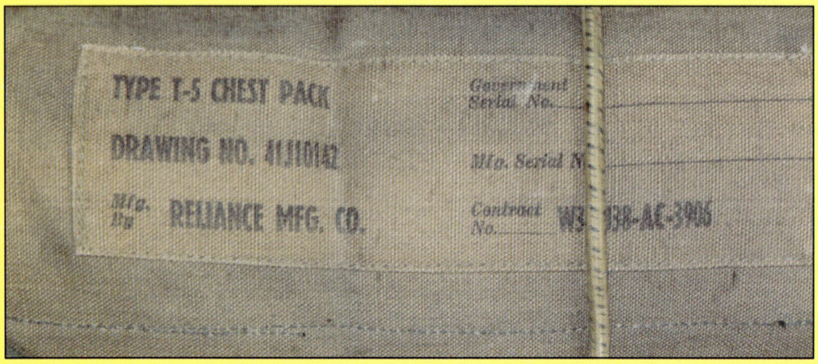

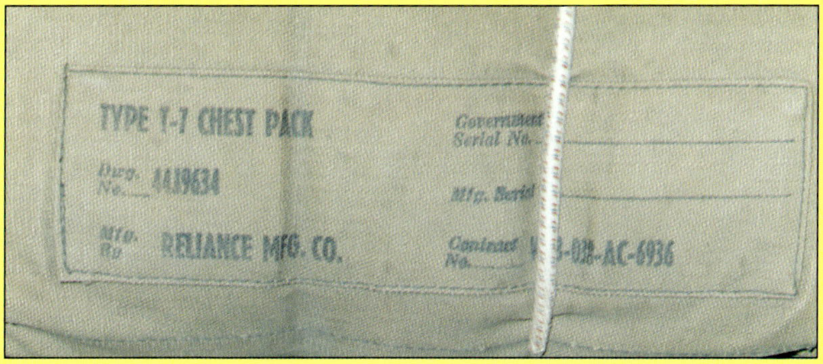

T7 reserve. No difference with T5A. Note 1945 date.

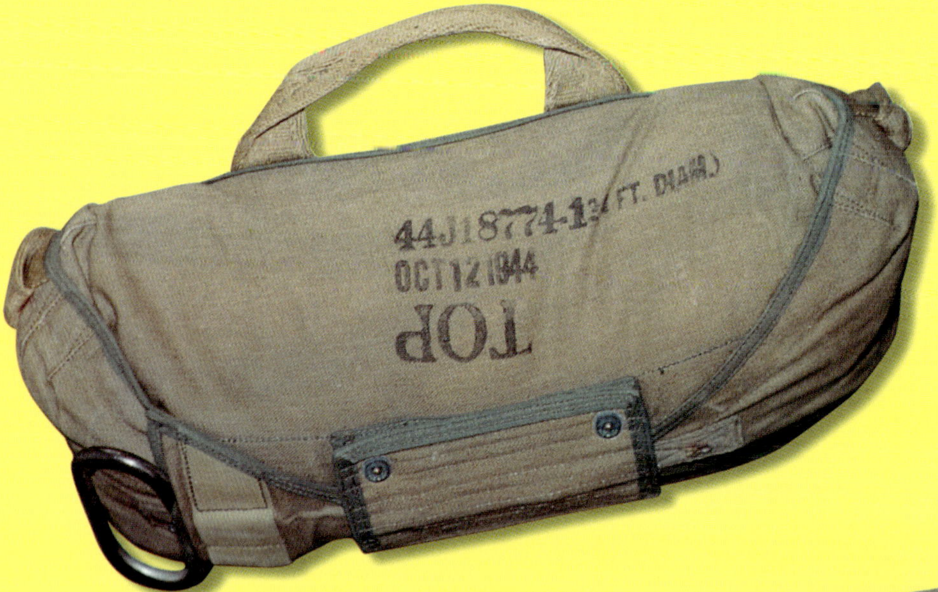

T5A reserve underside showing maker's data panel.

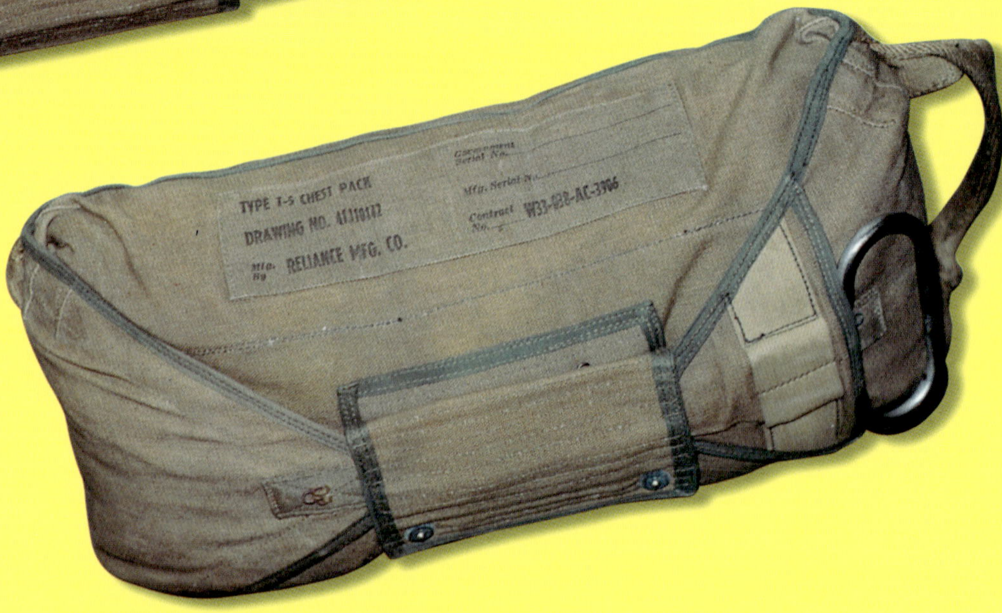

T5 main harness details. Notice the leg strap snaps are arranged in reverse order: right goes up, left runs down. This tray has unusual additional ties securing it to harness.

Late production T5A (1945). The T5 and T7 were both made of whatever color fabric was at hand. Thus T5As may be green, and T7s khaki-tan-brown.

Left: T7 in development, July 1942. This prototype was the YT7 made of undyed Type VIII cotton web. Note first model static line snap hook over wearer's left shoulder.

Top right: This photo perfectly illustrates the post-Normandy method of modifying the T5 harness to include the Quick Release box. The chest straps were removed and replaced with lengths of Type VIII webbing ending in QR lugs. The leg straps were retained, but snapped into short web adapters also of Type VIII webbing with QR lugs. These pieces were made in assembly line fashion on a grand scale in order to convert as many T5s as possible in the shortest amount of time. The T7 was a theatre expedient solution. The soldier on the left has donned his harness properly, while the trooper on the right has just snapped his leg straps directly into the QR box temporarily. He would not jump this parachute with the leg straps so misrouted. The airborne operation these troop 'chuted up for on August 17, 1944 was canceled.

Right: T5 modified to YT7 incorporating the Quick Release box.

2nd Model of T5 converted to YT7 configuration.

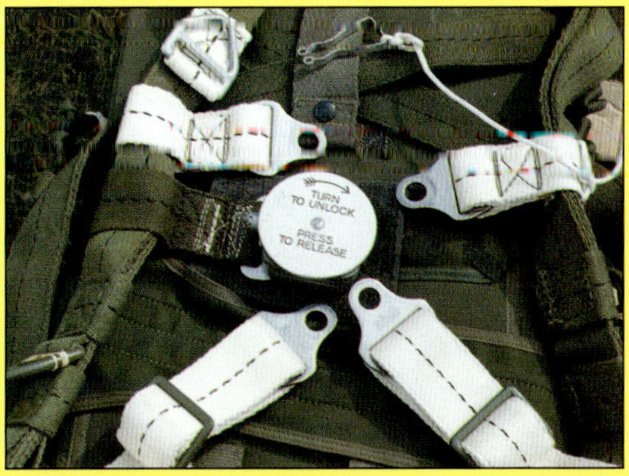

Above and below: Reproduction of the 2nd Model YT7, used by 17th Airborne Division in Operation Varsity. This repro is notional because load bearing webbing was not spliced.

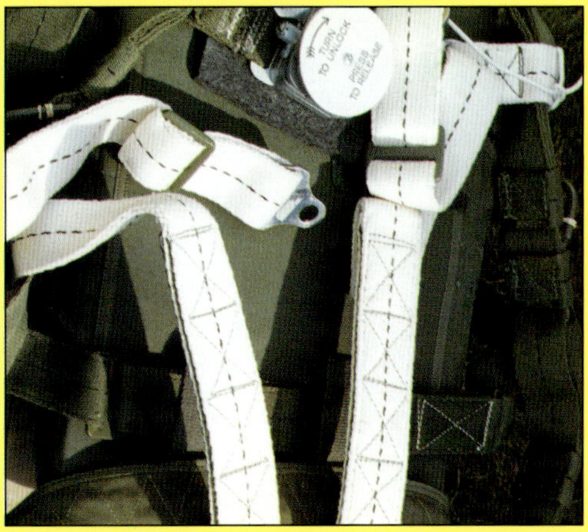

Below left: Quick Release Box released all four lugs and was retained by separate short strap. Below: Close-up of saddle showing how leg straps were anchored.

T5s modified to YT7 incorporating the Quick Release box.

WASHINGTON MERRY-GO-ROUND

Army Adopts Quick-Release Parachutes

By DREW PEARSON

Washington, June 23.

THE ARGUMENT over the quick-release parachute harness was revived this week by Life magazine, which contended that the U. S. Army preferred the slower triple-release harness which has to be unfastened over the shoulder and at each thigh.

This argument, however, was settled once and for all by Maj. General O. P. Echols of the Army Air Forces when he appeared before the House Appropriations Committee.

He was asked by Representative Case, South Dakota Republican, "What is the current policy with regard to the procurement of parachute equipment?"

General Echols replied: "The current policy is to procure the single-point quick release."

Representative Case: "Is the quick-release parachute regarded as having special value for flying over water?"

General Echols: "Yes."

Representative Case: "In other respects, is the other type of parachute regarded as useful?"

General Echols: "The quick-release single point, as I am now informed, has some advantages for paratroopers. They can release it with one hand and carry equipment with the other hand. It took them quite some time to arrive at the conclusion."

Most significant part of General Echols' testimony, however, was regarding the Navy.

"The Navy still has the policy of continuing procurement of the present conventional parachute," Echols explained. "They do not like the quick-release 'chute."

"It would be assumed that more of the Navy flying is done over water than the Army," Representative Case remarked.

"Yes, sir," replied Echols, and then explained that the Army had to break away from the Navy's policy of buying the old-fashioned three-point-release 'chutes. "The Navy did not agree," he said. "That had something to do with the slowness with which we arrived at a decision."

Questioned by Representative Case regarding comparative costs, General Echols replied: "The single-release parachute is more costly. But I am quite sure that the cost of the item itself, as compared with the importance of it, is very little and should not be given consideration."

• • •

Popular Senator Scott Lucas of Illinois received a phone call the other day from the secretary of Senator Kenneth Wherry of Nebraska, informing him that Wherry had a $1,000 contribution for Lucas' campaign.

Senator Lucas is up for re-election this year, naturally is receptive to any help from friends. However, he was surprised that he, a Democrat and stanch administration supporter, should get aid from Senator Wherry of Nebraska, a stanch Republican and bitter Roosevelt critic. Ever since Wherry replaced Senator George Norris, he has relentlessly gunned for Democrats.

So when Lucas encountered Wherry later on the Senate floor, he went up to him and said:

"Where's that thousand dollars you have for my campaign?"

"I don't have any thousand dollars for you," snapped Wherry in amazement.

"Why, your secretary called and said you had $1,000 for my campaign," declared Lucas. Then, noting the Nebraskan's discomfiture, he added: "Are you holding out on me?"

Flabbergasted, the Nebraska senator thought a moment and then cleared up the embarrassing mystery.

"That secretary of mine!" he exclaimed. "I told him to call up Rufus, meaning Rufus Holman (Senator from Oregon), and tell him I had $1,000 for his campaign. He must have thought I said Lucas and called you by mistake."

• • •

U. S. leadership in Latin America struck its lowest ebb this week when it had to bow before other Pan-American countries and promise to recognize the government of Bolivia, which Secretary Hull previously had ostracized.

Inside story of what happened is that Bolivia's neighbors long had been getting restless because of the "I-don't-speak-to-you" policy. Earlier they agreed to snub Bolivia, but with more and more signs of Bolivian-Allied co-operation, they failed to see the reason for prolonging the cold shoulder.

Finally, Brazilian Foreign Minister Oswaldo Aranha, now foremost, most forthright Pan-American leader, urged Secretary Hull to remove the ban. When Hull still delayed, Colombia, Brazil and Venezuela notified him that they would recognize Bolivia last Monday.

This put the United States on a serious hot-spot. Quickly, Hull wired his ambassadors to delay Brazilian-Colombian-Venezuelan recognition to the week-end, with the promise that they would then all go along together.

For ten years previously, it was the U. S. A. which took the Pan-American lead.—(c1944.)

DRAWING NUMBERS T4 and T5

Drawing numbers are the blueprints for each component and are provided here for identification. They have been replaced by Military Specification (MilSpec) numbers.

T4 system 41J1801, 41J1825[87]
T4 main canopy, 28' 065310
T4 pack tray 41G1824-1
T4 pack tray cover 09516-3
T4 reserve canopy, 22' 065311
T4 reserve container 069513-3
Fastener harness "bayonet", female 40A6465
Fastener harness "bayonet", male 40A6469

T5 main 41J10142, 42G2019
T5 & T7 28'canopy 42G2019-2, 42J2019, 42J2019-2
T5 reserve container 22' 065311, 41K10620, 42K4960-1
T5 harness 42J3710
T5 pack tray 41K9208-1, -2, -3
T5 pack tray frame wire 13 x 20" 40D6460
T5 pack tray cover with static line 41K9504-2
T5A reserve canopy 24' 42G2001-1, 42G2001-7, 43G14990 with A-1 pilot chute
T5A reserve canopy 24' pilot chute 046208 also type A-1A 43K21909 and A-1 45K9005
T5A reserve canopy 24' pilot chute camouflage fabric 69310-1, 69310-3
Frame, pilot parachute 48075 (two coil springs with long arms arranged cruciform to cause quick opening for the pilot chute.)
T5 reserve container 43H21008-1
T5 reserve (TR 1) 41K10620 22' canopy (no pilot chute)
Cord, pack opening (bungees) for reserve container 076177-4 (6 3/4"), 076177-3 (9 1/4")
T5 reserve ripcord assembly, rectangular handle 41B10625
T5 reserve ripcord assembly, cloverleaf handle 42B4964, 42B10910 ripcord length 6 3/4". Assembly
42B4964 has the pins wire wrapped and soldered to the cable. This method was superceded by press swaging pins into the cable strands for a much more secure attachment.
V-ring, parachute harness 43D21543
D-ring, parachute harness AN 6564, 43B21549
Links, connector AN6562-1, 0169131, 40A6470
Connector snap, chest pack 43A21544, 44A26449, 45A28113
Pin, static line safety 42A3519 for snap assembly 41B3674
Snap, parachute harness 41A3483
Snap, parachute anchor cable 41B3674 (static line snap hook) Used on T4 & T5.
Snap, parachute anchor cable 42B3232 ("hump back snap hook)
Snap, slide assembly parachute anchor cable 43A9502
Adapter, shoulder harness adjusting AN 6566 1, 40A6462
Adapter, harness strap AN 6565, 41A8728
Pack opening elastics 43A21914 These were made of rubber and would, in time, deteriorate. They were to be tested annually and had to stretch to 100% of length without damage. POE's were color coded for year and quarter on manufacture. A double red with single green thread woven in the cover meant 3rd quarter 1942. A single red with double blue equals 1st quarter 1943. (Spec.20-23 TO 04-5-2)
Chest parachutes AN6513-1, AN6513-1A, 42G14621 QAC types circa 1942 with 24' canopies that were otherwise identical to T5A chest packs.

36A6231 Fastener, parachute pack (on reserve end flaps)

36B6229 Cone, parachute pack releasing (reserve pack closure)

39A1875-1 Grommet, parachute pack
39A1875-2 Washer

40A6462 Adapter
40A6465 Fastener, harness
40A6468 Frame, wire harness
40A6469 Frame, wire harness
40A6470 Link connector

40D6460 T5 pack tray frame wire 13 x 20"

41A3483 Snap, parachute harness
41A8728 Adapter

41B3674 Snap assembly, parachute anchor cable
41B10625 T5 reserve ripcord handle, rectangular

41G1824-1 T4 pack tray

41J1801 T4 assemble, complete training type
41J1825 T4 assembly
41J10142 T5, complete training type
41J10142-2 T5 complete training type. T5 Troop Parachute, camouflage

41K9208-1, -2, -3 T5 & T7 pack tray
41K9504-2 T5 tray cover
41K10620 T5 reserve 22' canopy
41K10620 Pack assembly, parachute training chest (number refers to container or entire package)

42A3519 Pin, static line safety (screws into 41B3674)

42B3232 Snap hook, static line
42B4964 T5 reserve ripcord
42B10909 T5 reserve ripcord (3/8" shorter than 42B10910)
42B10910 T5 reserve ripcord
42B10913 T5 reserve ripcord
42B15006 Holder, rifle clip ("rigger pouches")
42B15007 Holder, sub machinegun magazines

42G2001-1, -7 T5 reserve canopy, 24'
42G2019 T5 & T7 main canopy, 28'
42G14621 T5 reserve container
42G15004 Holster assembly, parachutist's rifle (Griswold bag)
42G15005 Holster assembly, parachutist's SMG

42J2019 Canopy main 28', white
42J2019-1 Canopy main 28', orange & white, for arctic rescue unit

42J2019-2 Canopy main 28' camouflage
42J3710 T5 Harness

42K4960-1 T5 reserve container for 22' canopy, no carry loop

43A9502 Snap, static line (same as current MS 70120)
43A21538 Snap, parachute harness
43A21544 Connector snap, chest pack
43A21914, -2, -3 Pack opening elastics
43A21916 Pack opening elastics hooks

43B21543 V-ring
43B21549 D-ring (also 046201)

43G14990 T5 reserve canopy 24'

43H1008-1 T5 reserve container, 24' canopy, one carry loop on ripcord side

43K21909 T5 reserve pilot chute

44A26449 Connector snap, chest pack T5 & T7 ("butterfly" snaps)

44D18773 T5 reserve container, 24' canopy, 3 carry loops (also T7)

44J18774-1 T5 & T7 reserve container, 24' canopy
44J9634-3 T7 assembly

45A28113 Connector snap, chest pack

45K9005 Reserve pilot chute

There were five static line snap hooks used in the evolution of the T5

1. Snap hook 41B3674 (and possibly 0162075)
2. Snap hook 41b3674 with safety pin 42A3519
3. Snap, sliding gate 42B3232 the "humpback", weighs one pound
4. Snap, sliding gate 43A9502, has a truncated front edge
5. Snap, sliding gate 43A9502, is slab sided

None of the last three were pierced for safety wire. That was a post-war (1947) addition.

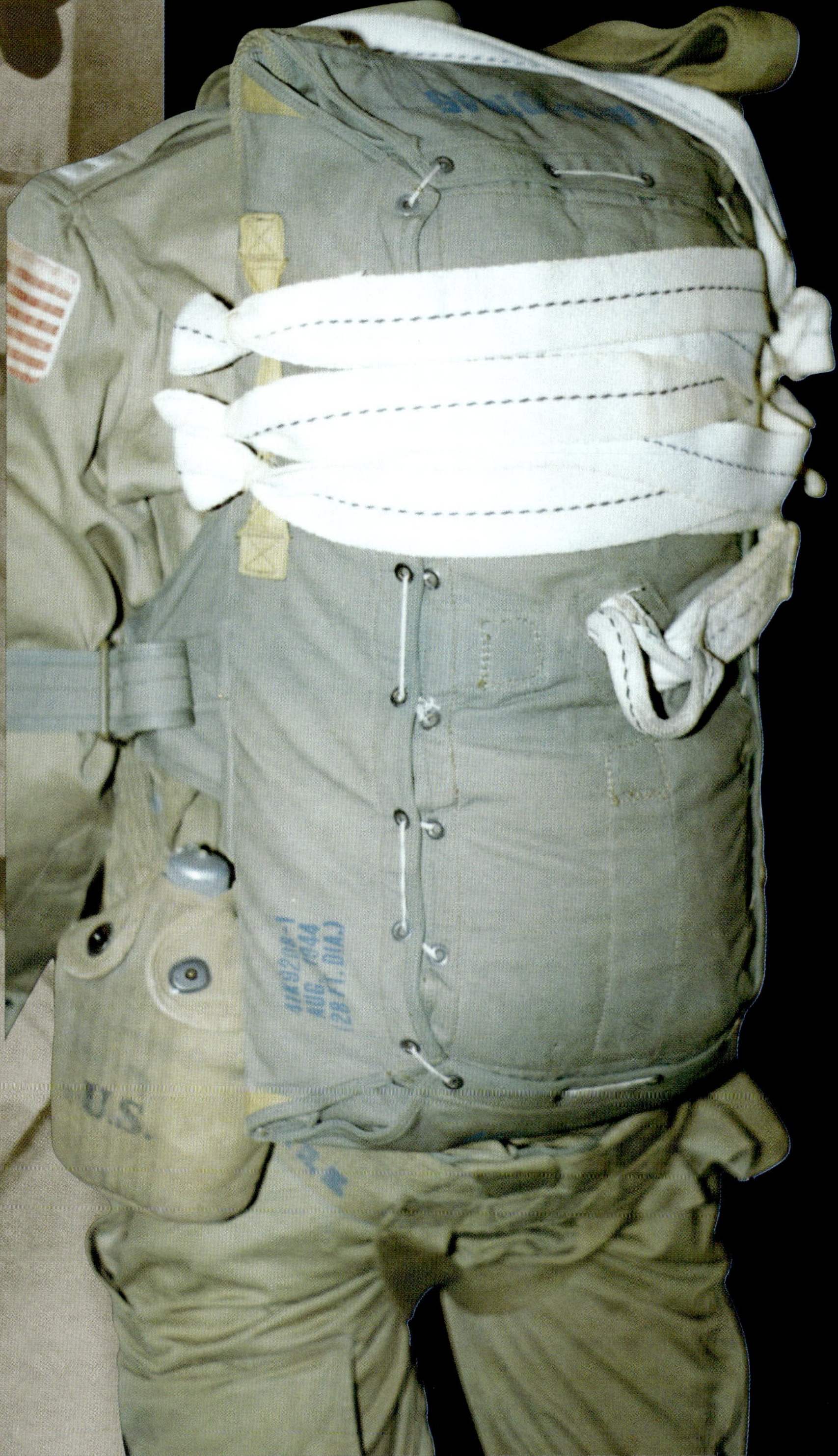

Perfect illustration of the pack tray cover being torn from the tray by the static line. Notice the canopy apex peak withdrawal as it is tied to the static line cover loop. The T5 and T7 functioned in the same manner.

T7: 24' chest reserve with pilot 'chute; single point harness Quick Release; harness made of Type X webbing; 4th or 5th Model static line snap hook, canopy camouflage material.

The Griswold bag attaching V-ring is substantially mounted. The Quick Release box safety fork is permanently installed. This tape may also be olive colored. The cords hanging from the shoulder adapters are a convenience in pulling slack in the webbing for adjustments.

Three photos at top: T7A is virtually identical to the T7. Note post-World War II date and pack opening spring bands.

Three photos at lower right: Operation of the Quick Release. The small M on the button signifies 'medium'. There was a plan to issue harnesses in three standard sizes – it was dropped as too awkward

TURN
TO UNLOCK
PRESS
TO RELEASE

TURN
TO UNLOCK
PRESS
TO RELEASE

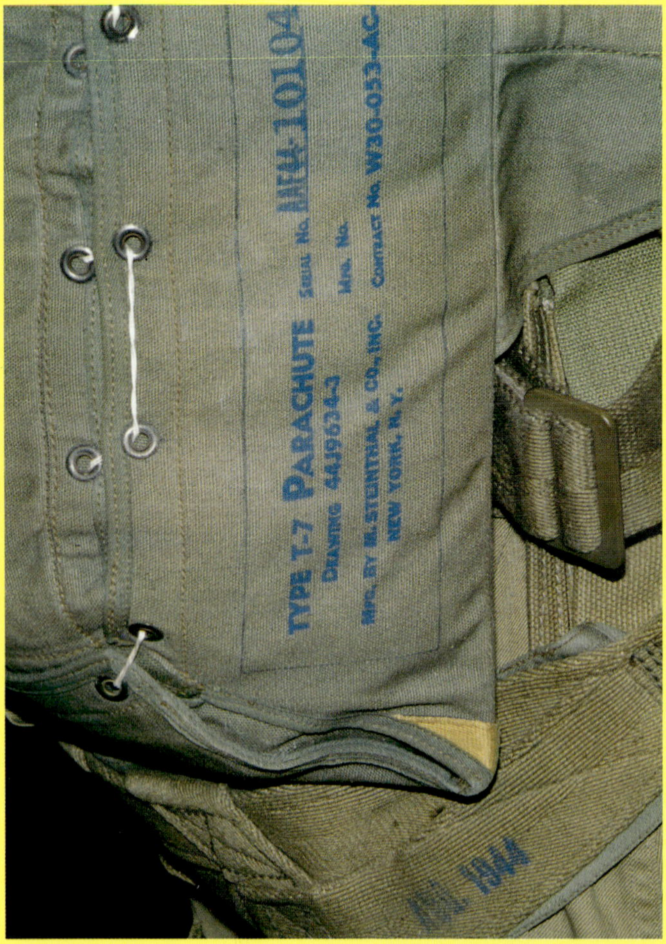
TYPE T-7 PARACHUTE
DRAWING 44J9634-3
MFG. BY M. STEINTHAL & CO., INC.
NEW YORK, N.Y.

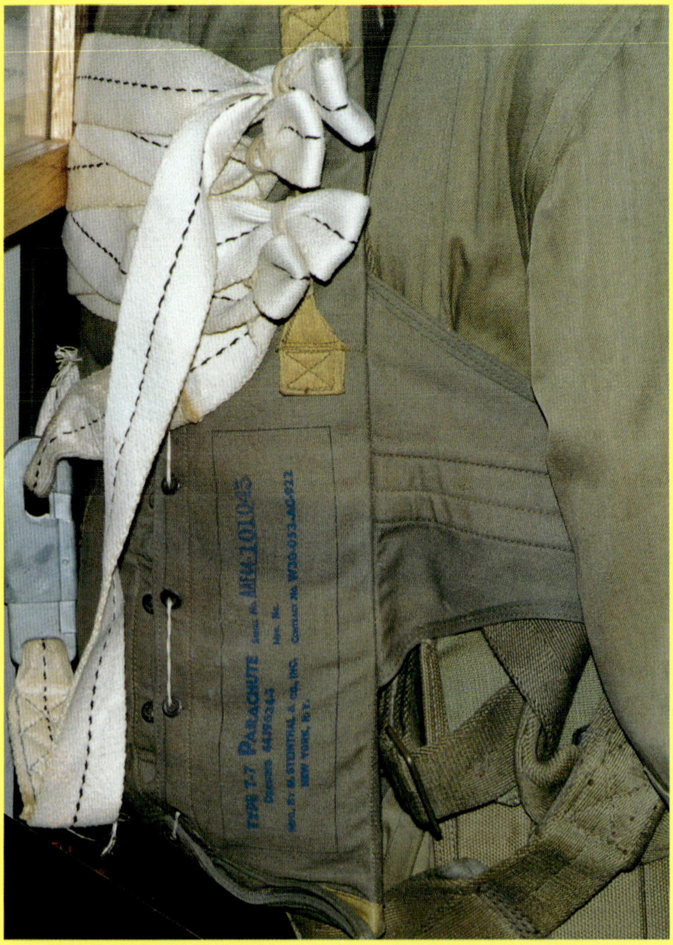

41K9208-1
AUG. 1944
(28 FT. DIA.)

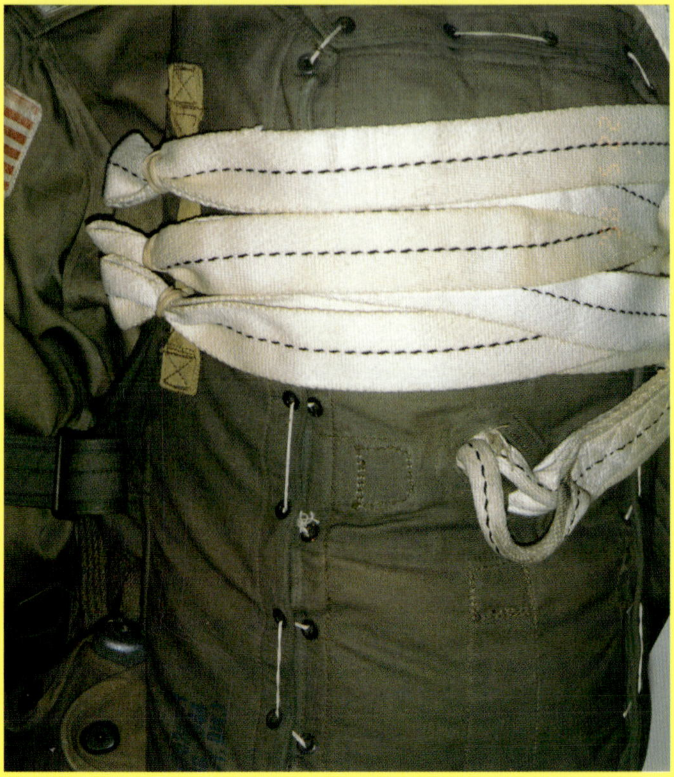

Opposite: The T5 and T7 were clearly assembled with whatever materials were at hand.

Right: Some veterans recall that reserve wear was made an option for Normandy and Holland, since they were to jump low. This is what they might have looked like, but with a T5 ... and of course more gear!

Late in the war the reserve snaps were improved by adding extended tabs. These "Butterfly" snaps made it easier to detach the bundle.

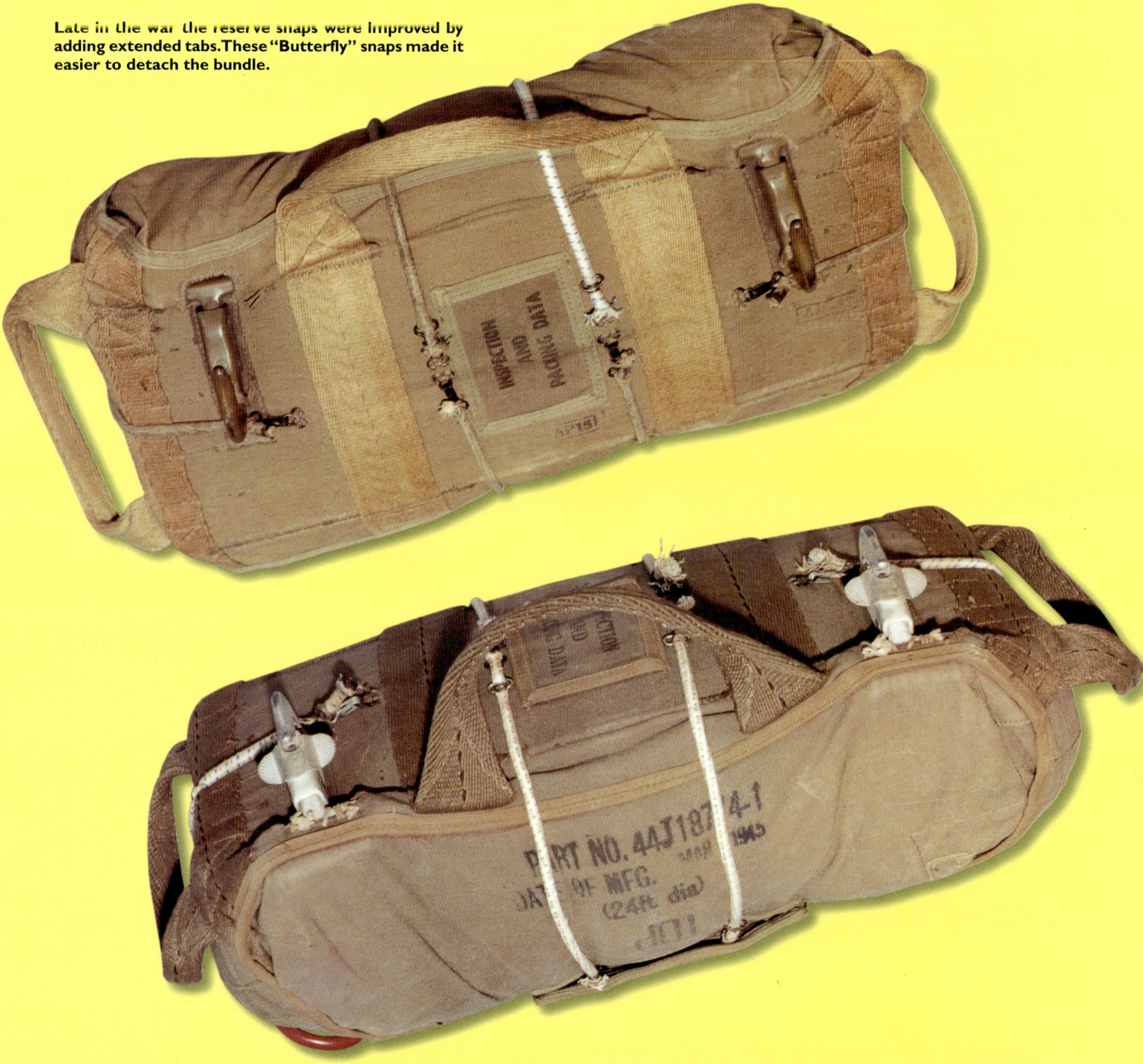

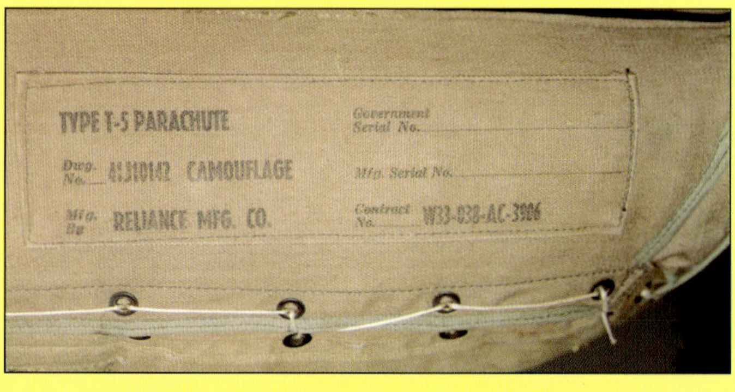

This data panel is noteworthy because it is a separate piece sewn on by Reliance. The Reliance conglomerate usually stamped their mark directly onto the pack tray presumably to save time.

These 17th Airborne Division troopers have donned the newly adopted YT7 main rigs prior to the drop across the Rhine. These early T7s were made of undyed Type VIII webbing. The red painted reserve handle was being phased out, and the leather chin cup was being replaced by the web strap.

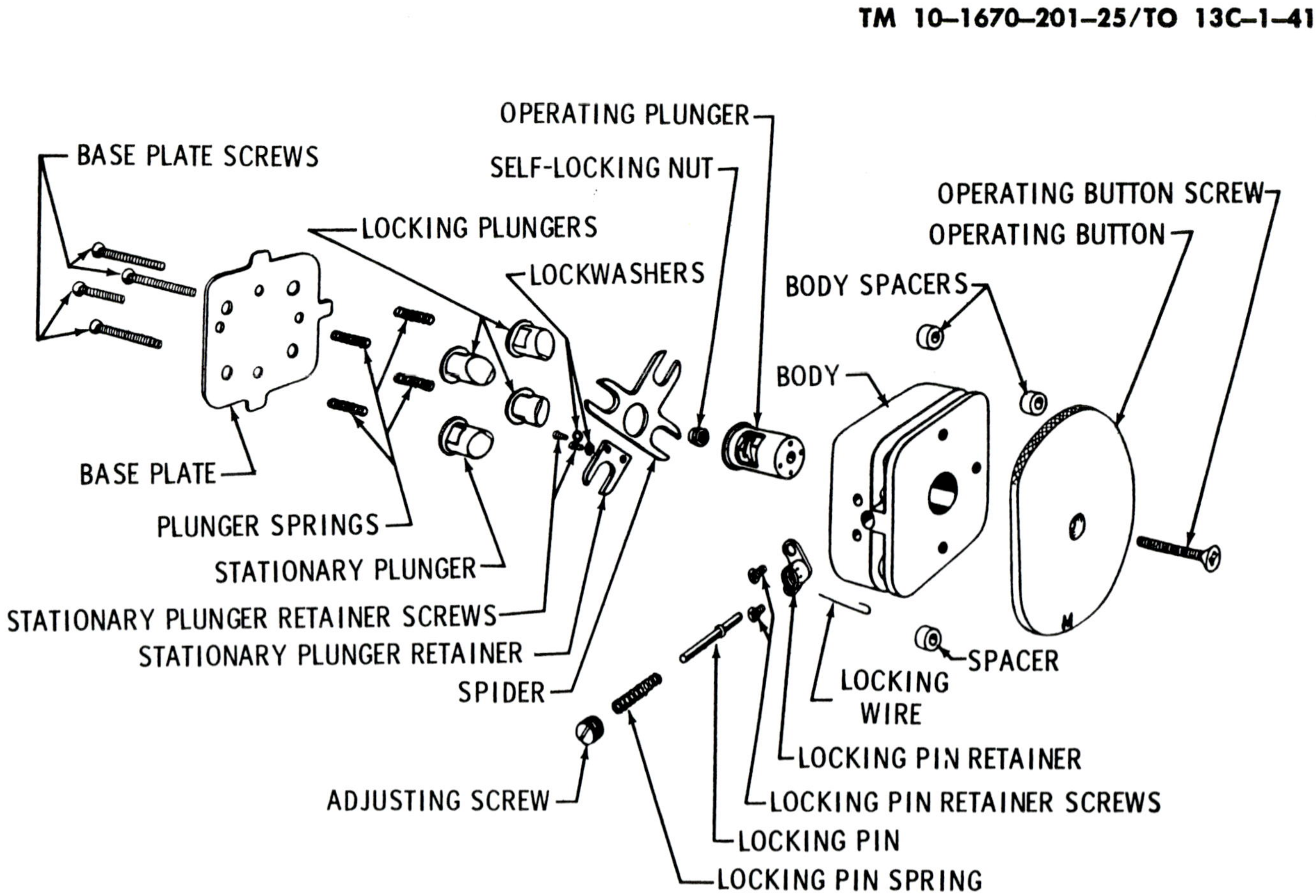

Figure 30. Parachute harness quick release, exploded view.

This postwar drop from a C-82 further illustrates the erratic T7 canopy deployment.

T7s will generally be found in olive green but they were also fabricated in the various shades of khaki/tan, mustard brown, olive drab, just as some T5s were made of green material. During the war years at least twenty-eight firms made parachutes or components and they used whatever approved material was supplied. There is a generalization that can be applied to fabric colors: they went from light khaki to mustard yellow to olive drab to olive green as the war progressed. The term "olive drab" is rather elastic and imprecise as thirty-three shades have been noted. Olive green shade 7 won the contest as the universal camouflage color by VJ-Day.

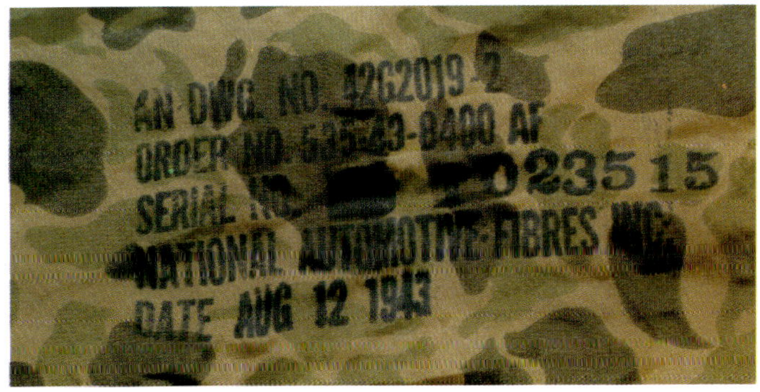

It may be impossible to determine now who made what, where, when and in what quantity as some of the wartime firms are many years out of business or combined with others. Sadly, some companies are know to have disposed of their archives.

T7 drawing numbers include:

T7 assembly 44D9634-3
T7 harness 44J9635
T7 canopy 42J2019
T7 canopy 44J9634
T7 canopy 48E7639-3 21' lines, post war
T7 pack tray 41K9208-3, 44J9636-0
T7 reserve container 45J18768
T7 reserve pilot chute 43K21909, 44G26459-1, 45K9005
T7 reserve ripcord assembly 42B10909, 42B10913
T7 reserve container 46R3086-1 (post war)
T7A reserve 48J7156 (adopted 1948)
T7A reserve canopy 48J7156 white twill, 16'10" lines
T7A reserve canopy 48J7156-1 camouflage twill, 16'10" lines
T7A reserve canopy 48J7156-2 white ripstop, 20' lines
T7A reserve canopy 48J7156-3 olive green ripstop, 20' lines
Reserve wire frame 42D14626 Reserve container 45J18768
Snap connector, chest pack 47D3138, 5000 lb "butterfly" snap
Quick release box safety clip 44AB9637
Quick release box 44D9334
Pad assembly, harness quick release 44B9347
Lug, harness quick release 44B9347
Button, release operating ("dial of death") 44B9338
Release assembly body 44B9335
Quick release assembly 44D9586, 44D9334, 45D18809

The T7 system was marketed commercially by Pioneer Parachute Co. for several years after the T10 was adopted. It was offered in a nylon harness with the static line stowed under cloth loops on top as the P-T-7 or P-TT-7. The P-TT-9 was the same but with a 30' canopy.

U.S. troops test British Statischute on the U.S.T4 harness. The first, second and very last wear a T4. Capt. Yarborough and Lt. Ryder went through the British jump school in 1941 and a small number of X types were bought from Irvin Industries, Canada and affixed to T4 harnesses. Tests proved that the bag deployed, lines first method was safer but the onset of war precluded adoption of a new model.

TACTICS AND TECHNIQUE OF AIR-BORNE TROOPS 134

FIGURE 22.—Suspended harness training. Parachutist pulling down sharply at moment of impact with ground.

U.S. landing technique before PLF.

Jumpmaster personnel inspection.

In the fall of 1943 tests were conducted at Laurinburg, NC airfield in the feasibility of dropping troops from gliders. Forty-two jumpers, eighteen from the C-47 and twelve from each glider could be dispatched.

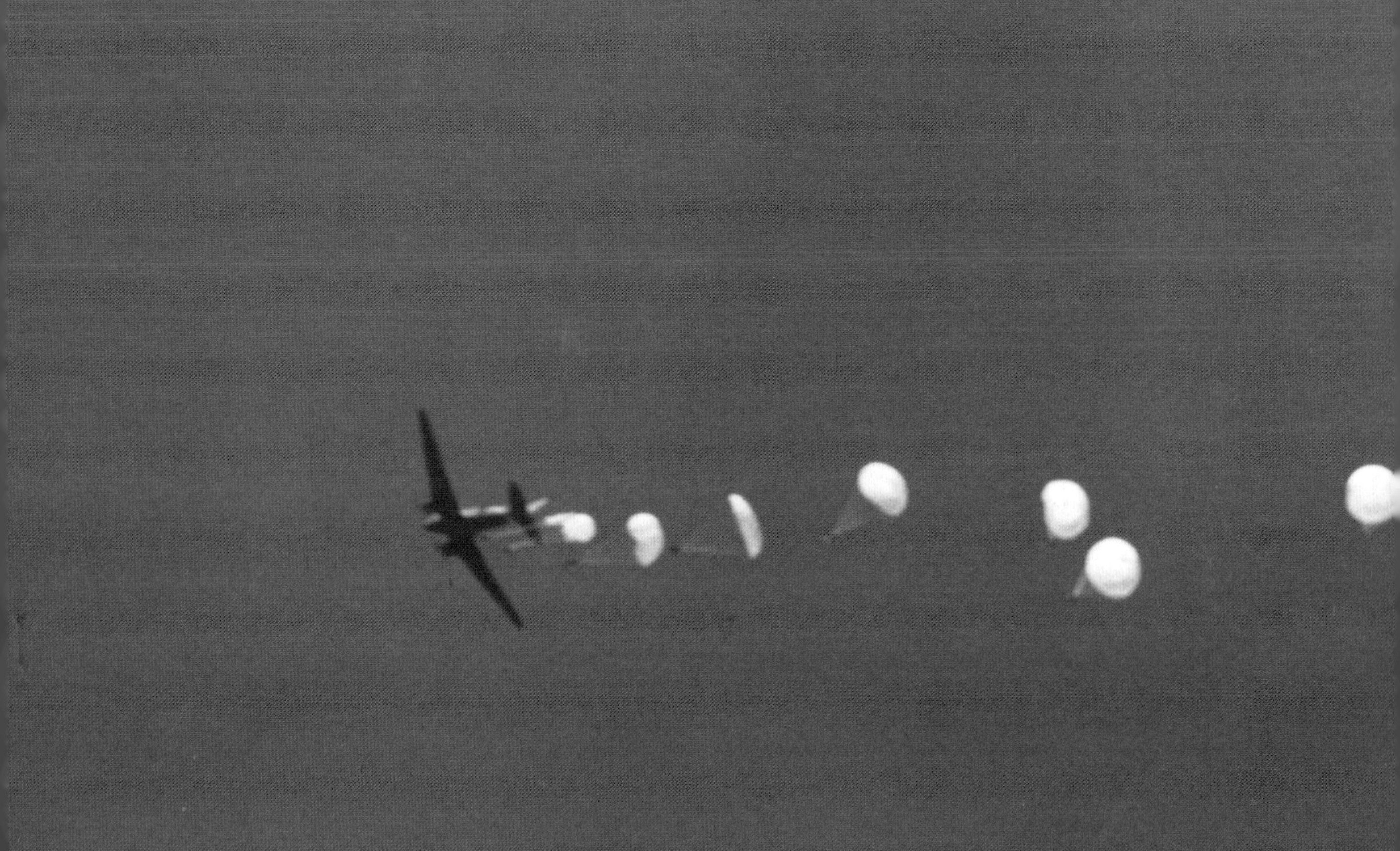

Canadian paras underwent the program at Ft. Benning before establishing their own school. Note the 3rd model snap hook 42B3232.

Comparison of current with wartime 4th model snap hook

Troopers exit both doors of a Curtiss Commando with the T5 modified.

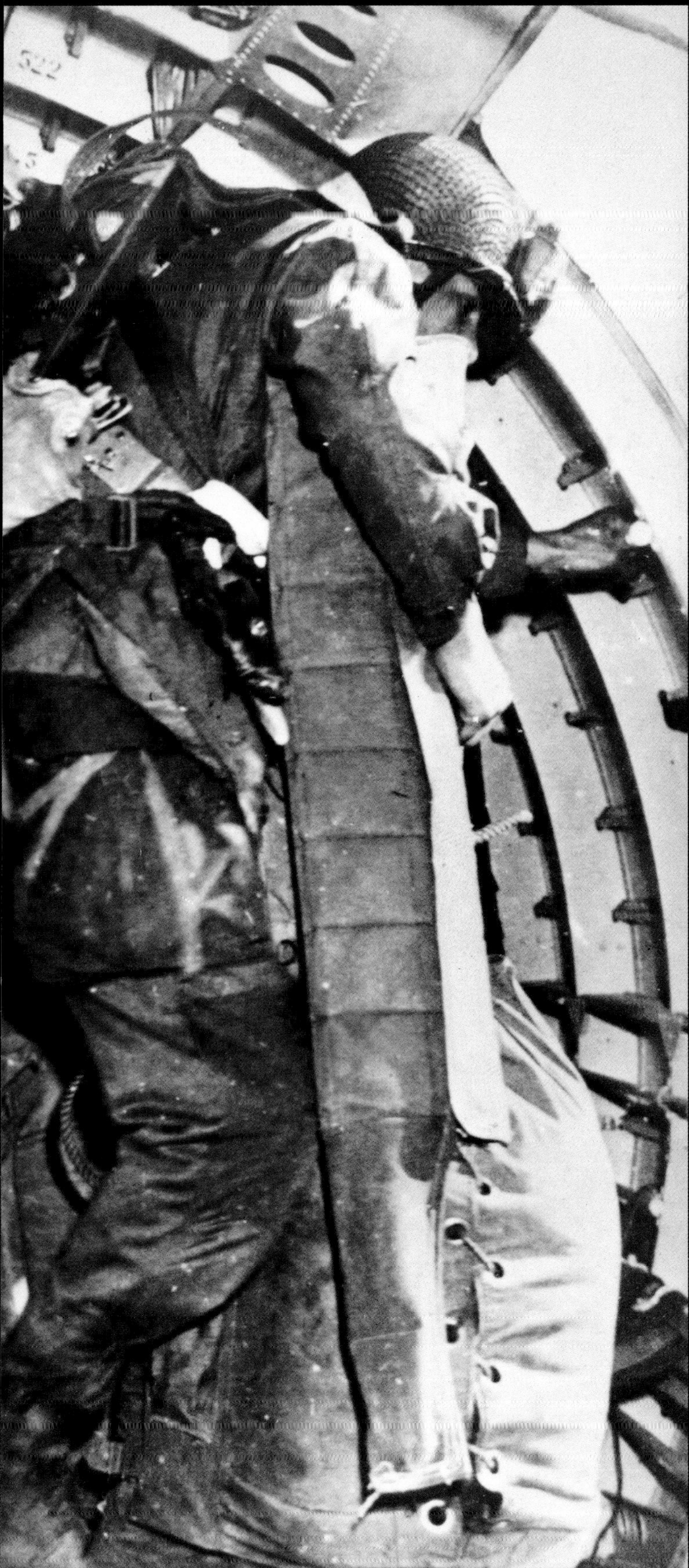

Clear illustration of how canopy T5/T7 could catch the skip stream and quickly break away.

This canopy is being withdrawn in a neat reefed manner – still connected to tray cover and static line.

Static line extended an instant before snatching off tray cover.

T5 and T7 developed in an unpredictable manner. The canopy was often blown back in a nylon bundle.

Here the tray cover is still attached to the apex. The canopy crown is filling like a sail and will snap the break cord.

Troops climb into a C-47 wearing T7s. France, March, 1945.

Troops leaving a C-46 with T7s.

This early T5 pack tray interior displays three rows of loops on which rubber retainer bands held suspension line stows. The apex pucker vent is also shown. This feature was deleted in 1942.

Packing the T5 or T7 Main

World War parachutes are not to be jumped, so this information is for those who wish to inspect, pack or repack their vintage rigs. It is included because the original instructions may be unclear at certain points.

No responsible rigger will repack a twill canopy and sign off on it as that is attesting that this parachute is jump worthy. Old twill fabric tears easily. Chances are good you will be doing this yourself.

Tools and supplies:

a Helper, not vital, but rather helpful
rubber bands, thick and sturdy type
a stout tie down cord for tying off apex to an anchor point
needle and thread
weights or shot bags
lacing cord, nylon twine will suffice
pack tray lacing instructions

1. Lay out on clean, smooth surface harness down, pack tray up. Tie apex to an anchor point

2. Straighten apex by pulling on both sides of the opening. Have helper pull back on risers or harness saddle to provide tension to the lines.

3. Separate the two riser groups. If the canopy lines are correctly fixed to the connector links, the data panel should be in the middle gore (bottom center).

4. Scoop up the front and rear gores in the middle. Drop all other lines, Separate front and rear gore lines. Run these lines back to the connector links. They should all be on the inside. This is a four line check and ensures line continuity all around the periphery. If the lines are snarled, and they often are, there are two ways to untangle. The easy way (and fun) is to take it out to a grassy field on a windy day and inflate it (assistance required). Then it is easy to see where to pass the risers/pack tray thru. The hard way is to try and figure what should go where when it is lying limp on the floor. It can be done!

5. Drop all but the front four on your side and pull out under all the lines. Pull out that entire gore. Pleat the canopy by pulling each suspension line up and whipping the fabric. Inspect the canopy during this procedure and note discrepancies.

6. Lay canopy down. Now you can see how tension helps. Count fourteen gores and lay them out so that the canopy looks like an isosceles triangle.

7. Dress up the canopy by gently pulling out any gores that are stuck in the wind channel. Don't bother with the topmost part.

8. Fold each side over to about the width of the pack tray. You'll need some weights to hold the material down. Riggers use shotbags for this purpose as they are ideal.

9. S-fold the canopy into pack tray lengths. Tie apex to cover loop with break cord. Of course, you must release the apex tie for this step.

10. Tack riser connector links to pack tray frame with needle and thread. Tie down riser tabs.

11. Begin stowing suspension lines. Make line bights over retainer band loops, then stow. Do not "milk" the lines. Make the stows uniform.

12. Place folded canopy on top of suspension lines.

13. Begin lacing the pack tray cover with about eight arm lengths of substitute cord. Study lacing diagram carefully in order to negotiate the corners correctly. It does not matter where you begin lacing.

14. Tighten lacing and tie off. Dress up pack with a short ruler or similar tool.

15. Stow static line across pack, starting with top right stow. Attach snap hook to center loop. There should be four stows on the right and three on the left.

16. Store in a cool, dry place. An Aviator's Kit Bag is a good receptacle, or a GI duffel bag will do just fine. Protect from sunlight, solvents, grease, oils and heat.

Parachutes should not be kept in a packed condition perpetually. They should be unpacked and aired out occasionally. Never in direct sunlight. Once a year is more than adequate. When you decide to sell it, be sure to get a signed statement from the buyer that he never intends for anyone to jump it.

Packing the reserve T5A & T7

1. Make sure the canopy is a proper fit for the container. A 24' canopy will destroy the 22' container.

2. Lay out on smooth, clean surface so that the back of container is up and tie off apex to an anchor point.

3. Straighten the apex and have a helper apply tension by pulling back on the lines or cross connector strap.

4. Separate the two line groups and scoop up the middle gore. You should see the data panel in the center.

5. Pull up front and rear gores. Place front gore lines in one hand, rear gore lines in the other and trace them back to the snap hooks. This is the line continuity check and the rear gore lines should run to the inside on the snap hooks.

6. Drop all but the rear gore line on the side you are on and pull it in and under all the other lines, then pull it out along with its entire gore.

7. Pleat the canopy. Grasp the S-line at the skirt and use it to fold the fabric over. Inspect each gore as this is being done.

8. The last line will have gores on either side. Look inside at the gathered lines, this is the wind channel. Ideally, it should be free of canopy fabric. It won't be, so clear it by spreading the lines or pulling back on the skirt. That will cause most of the material to fall out of the channel.

9. Lay the canopy down. It will look like a right triangle. Count off twelve gores (ten for a T5, 22') and lay out so that the canopy is now wedge shaped. Place weights to hold canopy down.

10. Flip container so that interior is up. Stow suspension lines by making bights over stow loops and slipping the rubber band over. Start stowing at lower ripcord side.

11. Fold canopy longitudinally into thirds. Remove the weights. S-fold into a stack with bottom slightly longer than the pack tray. Place on pack tray with skirt on ripcord side. Tuck apex under top fold. Don't fret if it all looks messy.

12. Route pull up cord thru cones and pull thru grommets. Close top and bottom flaps with temporary pins. Close ripcord side end flap and secure with ripcord pin. If your reserve is too big for the container, this is where you'll find that out. Forcing it can destroy the container. Consider stuffing it with nylon scrap for display.

13. Fold pilot chute and insert so that it is between the container flaps and the subflaps.

14. Close final end flap and secure with ripcord pin. Remove pullup cords. Dress up with a ruler or similar paddle tool. Reattach pack opening elastics or spring bands. Store with these disconnected. You may cheerfully punch the packed reserve to make it nice, round and symmetrical. A small piece of wire wrapped around the cable at the base of the terminal pin and thence looped about the end cone will serve to secure the container from being inadvertently dumped......and foil the inevitable joker who likes to see what happens.....when.....this (?).... is pulled . ("Ooops!")

Mannequins generally are not sturdy enough to support the weight of parachutes. You will be compelled to build a framework just to uphold the mass, if you are intent on a display.

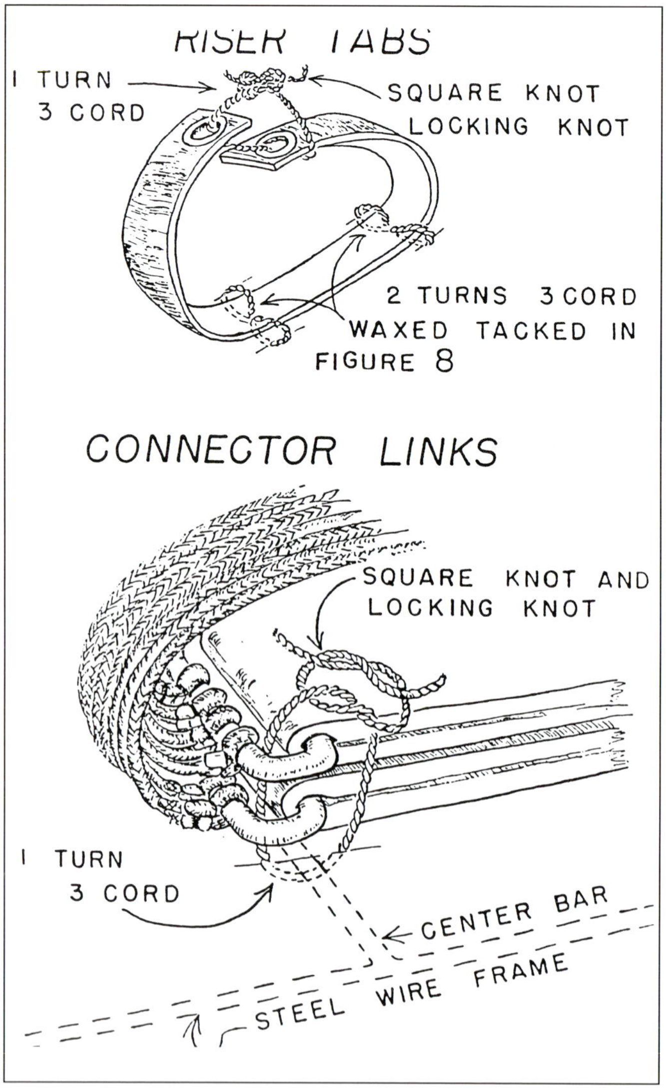

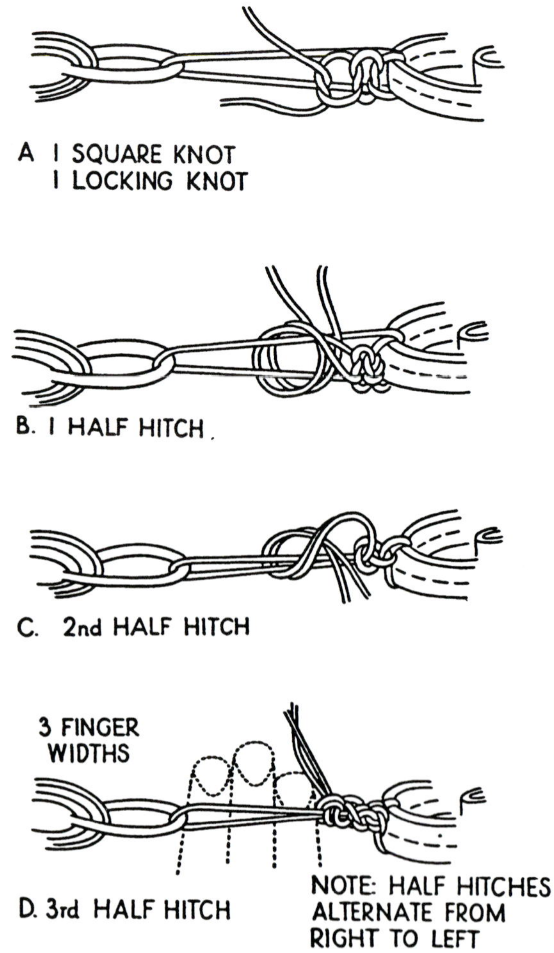

Figure 25. Tying the breakcord. (Text key: Par. 37b.)

Note. **For clarity one turn of 5 cord is shown.**

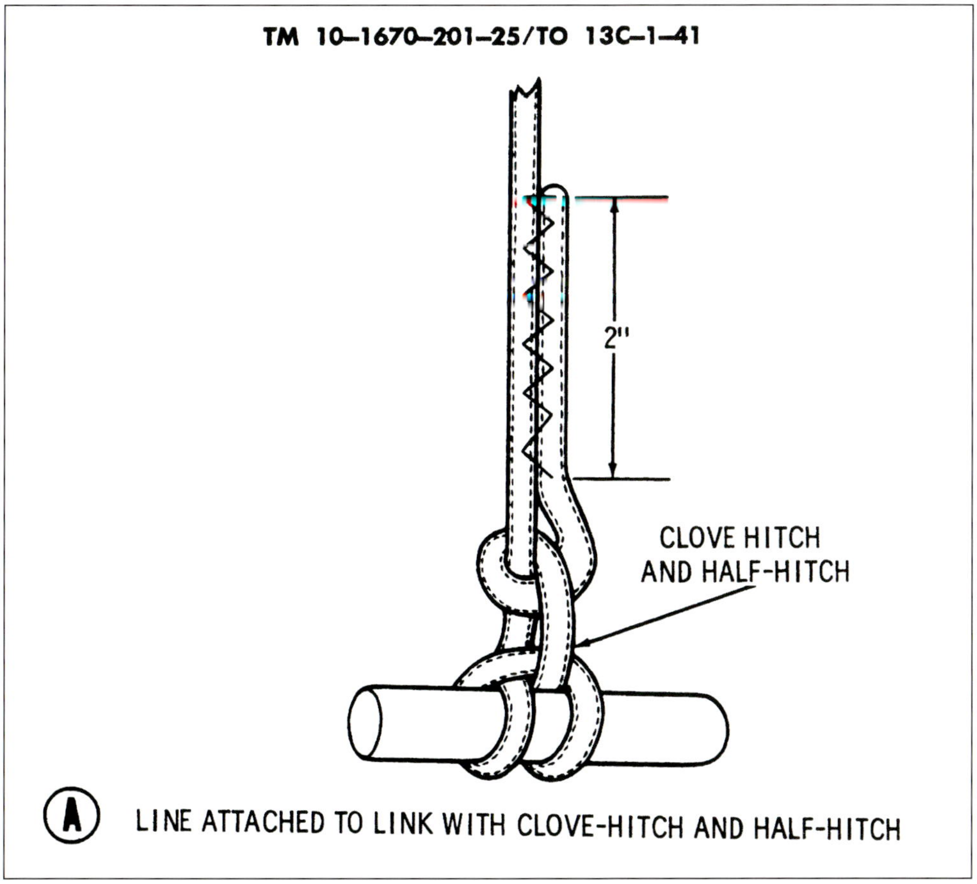
TM 10-1670-201-25/TO 13C-1-41
2"
CLOVE HITCH
AND HALF-HITCH
A
LINE ATTACHED TO LINK WITH CLOVE-HITCH AND HALF-HITCH

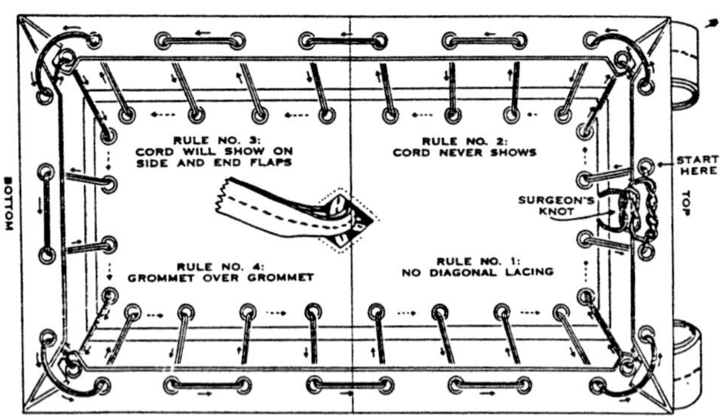

Figure 26. Lacing the pack. (Text key: Par. 37c.)

d. Stowing Static Line. The static line is stowed in the rubber retaining bands on each side of the pack. The No. 1 man folds the static lines for each stow at the edge of the steel wire frame of the pack tray. The first and last stows will be made on the right side. Four stows are made on the right side and three on the left. All retaining bands are doubled around the static line. (See fig. 27.) If the static line has been stowed correctly, it will extend from the last stow across the left shoulder to the finger tips of the left hand. A 3-inch loop of No. 5 cord through the left shoulder adapter is provided to hold the static line snap fastener.

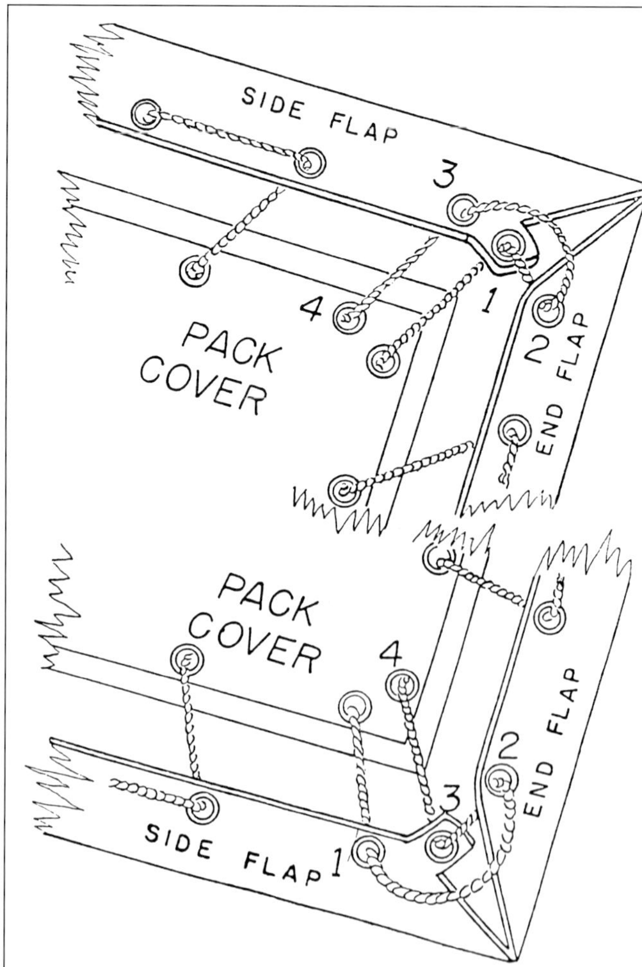

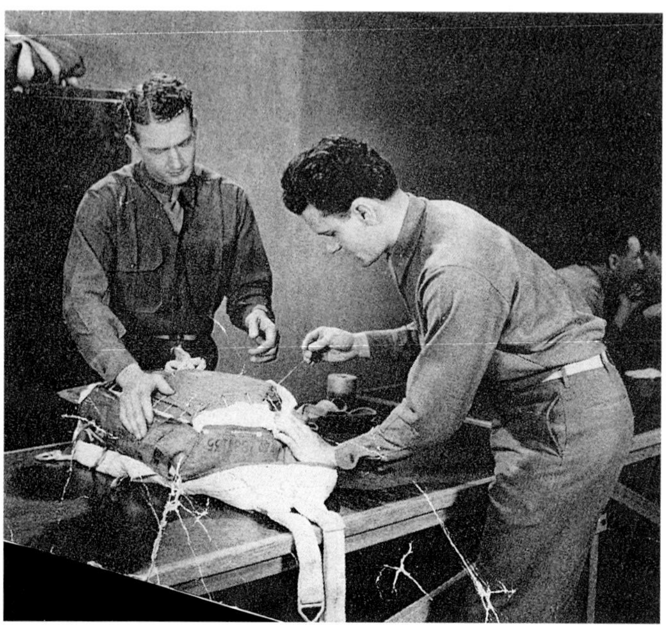

RESTRICTED

TECHNICAL ORDER
No. 13-5-7

HANDBOOK OF INSTRUCTIONS

FOR

Parachutes — Troop Type

13655

NOTE: This Technical Order replaces T. O. No. 13-5-7 dated February 14, 1942.

NOTICE: This document contains information affecting the National Defense of the United States within the meaning of the Espionage Act (U. S. C. 50:31:32). The transmission of this document or the revelation of its contents in any manner to an unauthorized person is prohibited.

PUBLISHED BY AUTHORITY

OF

THE COMMANDING GENERAL, ARMY AIR FORCES

BY

THE AIR SERVICE COMMAND

AIR FORCE SECTION

WRIGHT FIELD

DAYTON, OHIO

SEPTEMBER 1, 1942

FOTO-LITH, INC.

8-31-42—8000

TABLE OF CONTENTS

LIST OF ILLUSTRATIONS

RESTRICTED T. O. No. 13-5-7

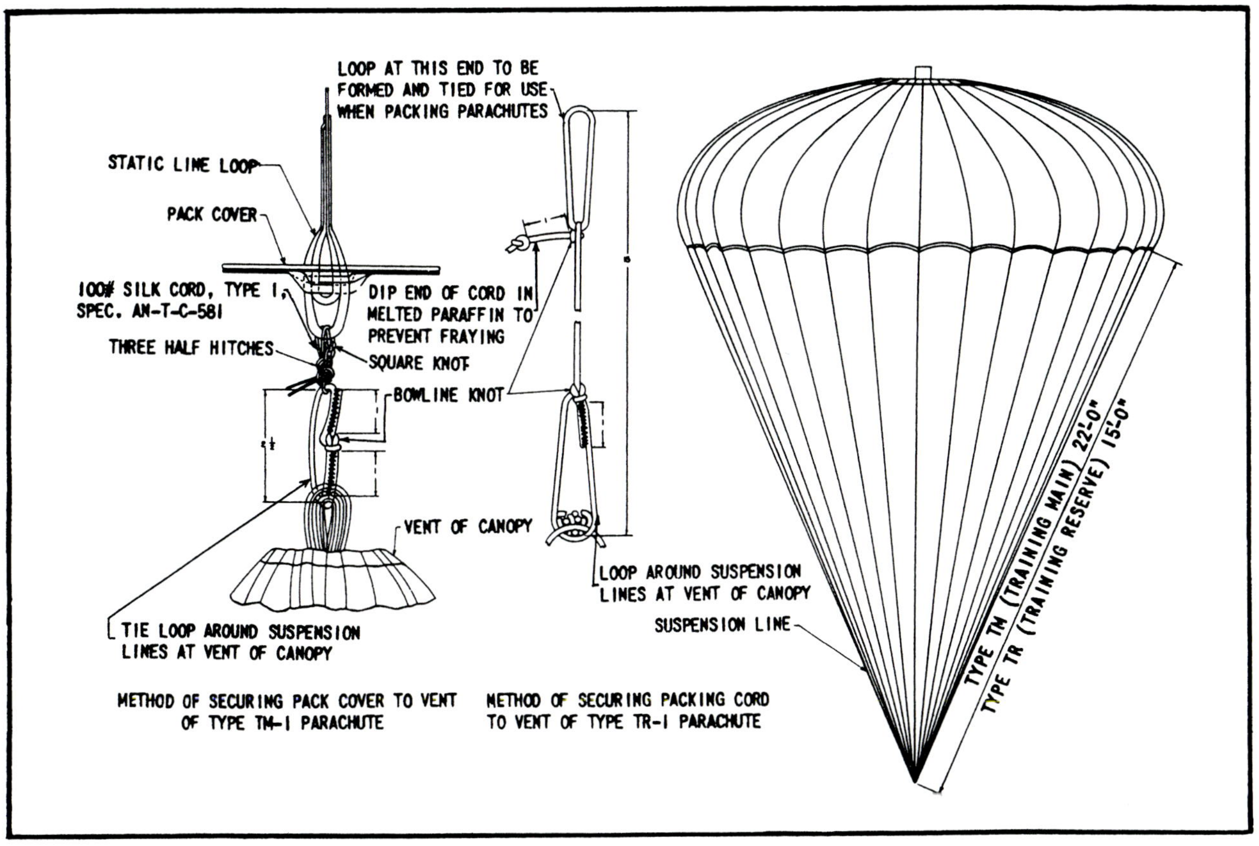

Figure 1 - Cording

SECTION I

INTRODUCTION

1. This Handbook of Instructions covers the semi-fixed station type, man-carrying parachute, which is designed for use in landing personnel on terrain inaccessible to aircraft and only when the jumps are premeditated. This type of parachute is actuated by means of a static line attached to the airplane. The parachute assembly consists of three main parts: main pack, reserve pack, and harness. The main pack and reserve pack assemblies incorporate conventional, circular-type canopies. The data below pertain to this type of parachute assembly:

2. The recording and reporting of drop-tests, inspections and repacking, as prescribed in the following sections, will be accomplished on A.A.F. Form No. 46, "Parachute Record", in accordance with the instructions in Air Corps Circular 15-46. A record of inspections and drop-tests will also be made on A.A.F. Form No. 58, "Parachute Inspection and Drop-Test Card", which will be retained in the pack assembly pocket of each parachute.

Type	Dwg. No.	Description	Diam. in Ft.	Wt. of Wearer
T-5	41J10142	Parachute, Training, Main Back	28	No limitation
		Parachute, Reserve, Chest	22	No limitation

SECTION II

DESCRIPTION

1. Canopy.

a. Construction.

(1) The canopy, which is constructed of high-grade silk fabric, constitutes the lifting surface of the parachute assembly, and has a framework of cords, or lines, known as suspension lines by which the load on the canopy is suspended. (See figures 1 and 2.) These lines vary in number, depending upon the size of the canopy. Each line extends from a connector link on one side of the harness riser to the skirt of the canopy, thence through a channel in the silk covering, entirely across the surface of the canopy to a point on the skirt diametrically opposite, and thence to the riser on the other side of the harness. This permits each line to be a continuous length from one side of the harness, up to and entirely across or over the canopy, and down to the opposite side of the harness. The lines cross each other at the apex of the canopy.

(2) That part of the canopy between any two suspension lines is known as a panel.

(3) The panels are further subdivided into sections. The silk of each section is cut on a bias, with the warp of the fabric running at a forty-five degree angle to the center line of the panel. This arrangement not only permits greater economy in cutting the fabric, but also prevents any runs or tears entending beyond the seams that are formed by joining the sections together. (See figure 2.)

(4) Completely encircling the panels at both the skirt and apex of the canopy are bands of 300-pound reinforcing silk tape. These are known as lateral bands, and serve to reinforce and strengthen the body of the canopy.

(5) A device at the apex opening, known as the puckered vent, functions to relieve the initial opening strains on the canopy. This operates by means of an elastic, which permits the vent to open wide under excessive opening pressure caused by high speed, and then closes as the parachute reaches the normal rate of descent.

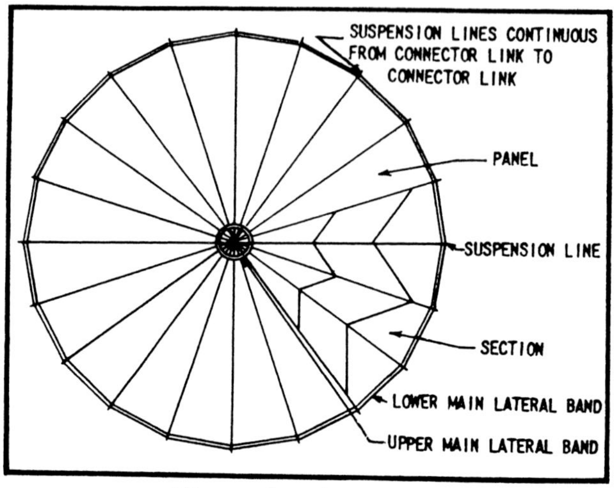

Figure 2 - Canopy

b. Types of Canopies.

Canopy Type	Dwg. No.	Used With	Canopy Diam. in Ft.	Rate of Descent Ft./Sec.
TM-1	065310	Parachute, Training, Main Type T-5	28	14 to 20
TR-1	065311	Parachute, Training, Reserve, Type T-5	22	17 to 26

2. Pack.

a. Construction.

(1) The pack is that part of the parachute assembly in which the canopy is packed and carried. The bottom of the pack is semi-rigid in construction, being reinforced by a wire frame made of solid drawn wire. An inner bottom, containing loops made of rubber, is provided for holding the folded suspension lines. The sides and ends of the pack consist of flaps fabricated so as to cover the sides and part of the top of the folded canopy. The cover, which is rectangular in shape, is laced to the pack flaps with two turns of cotton thread, Spec. AN-V-T-279, passing continuously all around the pack through the grommets located in the cover and the side and end flaps. The cover is fastened to the canopy apex with one complete loop of 100-pound silk cord. At the outside of the cover, a static line is attached at the center as a permanent part of the cover. This static line is fifteen feet long and is made of 3000-pound webbing. The free end of the static line has a snap fastener for attachment to a suitable anchorage within the airplane when jumping.

(2) The reserve pack is a conventional frame and fabric type. The frame portion has been reduced to a minimum to make the pack soft. The pack has the standard flaps equipped with quick-opening rubber cords and is released by means of the usual rip cord arrangement. Two fabric loops are provided at the underside to provide retainers for the belt to pass through. Snaps connected at the canopy riser ends pass through slits in the back of the pack for quick attachment to V-rings at the parachute harness.

b. Types of Pack Assemblies. - The following are the types of packs used:

Type	Drawing	Used With
Back - Main	41K9208	Parachute TM-1
Chest - Reserve	41K10620	Parachute TR-1

3. Harness.

a. Construction . - The harness consists of a webbing framework that secures the parachute to the wearer. The principal element is the main lift web, which passes from the point at which the canopy is attached around under the seat of the wearer, forming a swing. The webbing area is broadened at the seat, at which point the leg webbings are attached. An adapter is positioned at the front of the wearer's shoulder. Double back straps are passed around the center bar of the shoulder adapter, over the shoulders across the back through a loop in the main lift web to a snap or V-ring at the lap of the wearer, back through the lap of the wearer, and then back through the lift web across the back to adjustment adapters. The chest straps are located at the chest of the wearer twelve inches below the chin and are connected by a snap and V-ring assembly. The leg straps are attached at the seat sling and brought up between the legs and connected to the snap and V-ring fittings at the front loop of the back strap. The harness is made of a flexible cotton webbing. All webbing is doubled, and sewed joints are kept at a minimum. Single webbing strength is 3000 pounds. A total of eight webbing straps are included in the riser assembly. V-rings are provided at the front mid-section of the wearer for attachment of the reserve parachute. Adjustment is provided in the leg and back straps.

b. Types of Harness.

Dwg. No.	Used With	Parachute Assemblies
42J3710	Training Type	T-5

4. Flyer's Kit Bag.

As a convenient means for carrying and transporting the parachute when it is not being worn, there has been provided for issue a suitable container, Flyer's Kit Bag, type A-3, part No. 0158791. This bag is constructed of 12-ounce, double-filled duck reinforced with cotton webbing.

SECTION III

INSPECTION

1. Routine Inspection.

All parachutes that are issued for service will be checked for general condition and serviceability at least once every ten (10) days, and a record of this inspection will be maintained on Form No. 58. This will be a responsibility of the operating activity or organization to which the parachutes are issued. The inspector assigned should be a graduate of the Parachute Course at the Air Corps Technical School, but if such a graduate is not available, the inspector should be carefully selected, trained and approved by the Officer in Charge of the local parachute section. Accomplishing and recording a ten (10) day inspection of parachutes being used by personnel on missions for a protracted period is the responsibility of the pilot in command of the aircraft or his authorized representative. At activities maintaining a pool of parachutes at operations, inspection of Form No. 58 will be made prior to the issue of each parachute, to determine if the inspections required have been accomplished at the prescribed periods. These checks, which will be as complete and thorough as possible without breaking the seal and opening the pack assembly, will be made as follows:

a. Note the date of last inspection and repacking, also that of the last drop-test, as entered on the Parachute Inspection and Drop-Test Card, Form No. 58. If the specified period of these operations has elapsed (see paragraph 1., section IV and paragraph 2., section V), the parachute will be withdrawn from service for the accomplishment of the necessary work.

b. Carefully observe the external condition of the pack, harness assembly and static line assembly for any visible defects or deterioration, protruding silk at the corners, and for any acid or oil stains likely to cause deterioration of the contents of the pack. Note particularly the condition of all stitching. Check the harness and static line webbing for any damaged or weak spots, and for any rusted or defective fittings or snaps.

c. Check the condition, elasticity, and proper attachment of the pack-opening elastics.

d. Open the rip cord flap, and note whether or not the rip cord prongs are properly inserted in the cones. Check also to see that the prongs are not corroded or bent, and that the rip cord seal is intact and properly applied.

2. Disposition if Defective or in Doubtful Condition.

If, as a result of the foregoing check, conditions are found that indicate need for repairs or a more thorough inspection, the parachute will be turned over to the parachute department for unpacking and complete inspection before further use, irrespective of the date of last repacking.

3. Complete Inspection.

To prolong the life and maintain the reliability of parachutes, they will be unpacked, carefully and thoroughly inspected, and repaired as found necessary, at the intervals specified in paragraph 1., section IV. To accomplish this inspection, unpack and stretch the parachute out on the packing table, securing it at points near the apex and just below the suspension line connector links so that it will be held taut. This can be accomplished by the use of removable "pegs" placed in the parachute packing table, to which the various parachute fittings or lines can be attached as required, or by the use of weights. The harness will be placed in a position that it would occupy if its wearer were lying on the table face downward, with his head toward the top of the parachute. (See figure 7.) Separate the two respective groups of suspension lines as illustrated. Then proceed with the inspection, as follows:

a. The silk of the canopy will be inspected for damaged and weak spots by carefully looking over each panel and all seams. Mildew, rust, water, and oil stains, battery acid, and so forth, will cause silk fabrics to deteriorate rapidly. If evidence of any of these conditions exists, the necessary repairs will be made, after which if there is any doubt as to the serviceability, the parachute will be drop-tested or reported for overhaul. The condition of the silk will not be tested by attempting to tear it. For method of patching canopy, see figure 3.

b. The rubber vent ring of the canopy will be inspected and replaced if deterioration is noticeable, or if broken. To replace, use a round moulded-rubber strip, stocked as strip, rubber, extruded, type II, Spec. 20-82, class 04-B, of sufficient length to thread through the vent (approximately 18 inches). Splice the ends together so that the opening of the vent is four inches in diameter, with no appreciable tension in the rubber. The splice is made by lapping the ends of the rubber approximately one inch and wrapping with one turn of adhesive tape. The adhesive tape is then wrapped with thread. (Cotton, high tenacity, Spec. AN-V-T-279.) The function of the adhesive tape is to prevent the thread from cutting the rubber.

c. Each suspension line will be visually inspected, and if evidence of deterioration is found, the parachute will be reported for overhaul. The point at which the suspension lines are attached to the connector links on the harness will be particularly observed for evidence of rust. Under no conditions will suspension lines be repaired by means of knots or

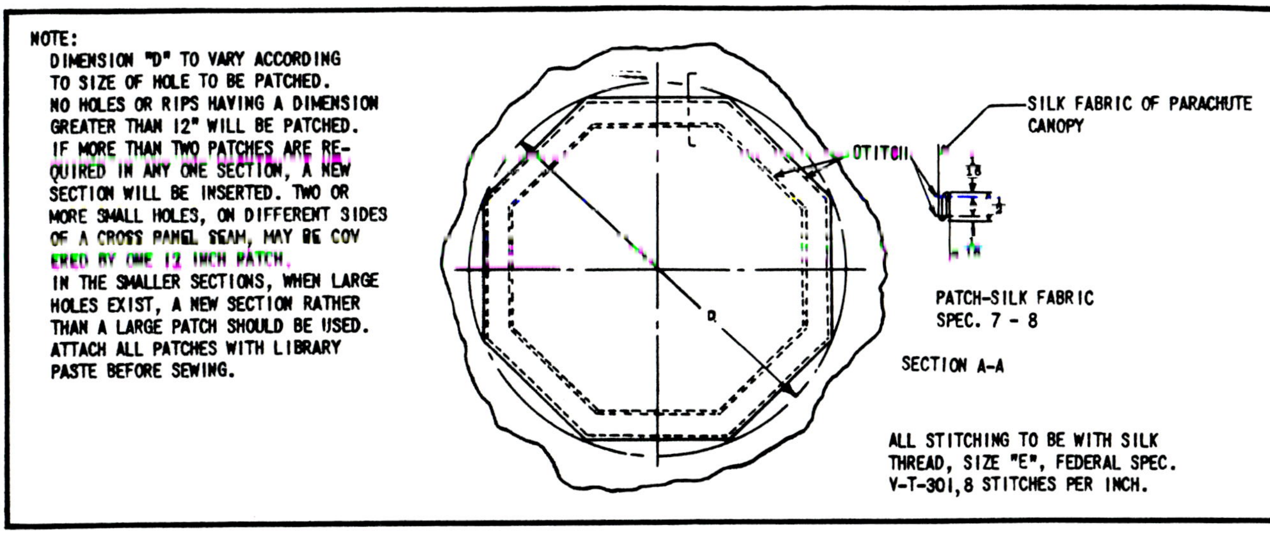

Figure 3 - Patching Canopy

splices. In the event any of the lines have been pulled loose from the zigzag stitching at the lower main lateral band, the parachute will be reported to the control depot.

d. The pack assembly will be carefully inspected visually for any defects or deterioration due to wear and tear in service. Pack assemblies in which holes have been worn or which show other signs of deterioration will be replaced, except that where the condition of the pack warrants, minor repairs that can be made by stitching, or patching small holes, may be accomplished locally if the necessary facilities are available. However, under no circumstances will stitching or patching be accomplished while the parachute is packed. Bent or broken frames that cannot be repaired locally will also be replaced. Stains from water, oil, and so forth, which may cause deterioration to the silk, will be removed if possible by washing the discolored area with warm water and soap, white, floating, Spec. No. P-S-616 (Ivory). Care will be taken to remove the traces of soap. The afore-mentioned soap, or equal, if not on hand, will be procured locally.

(1) Grommets will be inspected and replaced if no longer serviceable.

(2) The cones in which the rip cord prongs are inserted will be carefully examined and all rough places and sharp corners removed. The surface of these cones should be perfectly smooth.

(3) Pack-opening elastics will be replaced whenever any indication of deterioration or loss in elasticity is observed. When installing elastics, the hooks on each end will be inserted in the eyelets on the pack with the open part toward the pack, and clinched after packing is completed. If the end of a hook is allowed to remain open, it may wear a hole through the pack.

(4) Inspect the rip cord to make sure that there are no rust spots and that the wrapping and soldering on the pins is smooth and unbroken, and that the pins are not bent or defective in any other manner. Rip cord assembly, part No. 41B10625, will be inspected for soft locking pins. This inspection will be made by inserting the end of the rip cord pin one-half inch between the soft metal lined jaws of a vise, or in a jig made by drilling a hole one-half inch deep in a soft metal block. A five pound weight will then be hooked to the rip cord grip, care being taken to apply the load gradually without impact. Care will also be taken to insure that the surface of the pin is not marred in any manner. As a result of this inspection, all rip cord assemblies that show evidence of the pins having taken a permanent set will be reworked by the installation of a new cable and pins.

(5) Under no circumstances will rip cord assemblies be continued in service in a condition that might create a hazard if their use should be required in an emergency. This applies particularly to the condition of the cord at a point adjacent to the grip, below the solder, at which point the flexing of the cable tends to break the strands. To provide against the possibility of failure, all rip cords will be carefully inspected for broken cable strands, particularly adjacent to the grip, and for any other defects that might indicate failure of the assembly in normal use.

(6) The rip cord ring pocket must hold the ring securely and permit the rip cord grip to protrude sufficiently to make it instantly accessible. Any pocket that allows the grip to fall out, or which is so tight that withdrawal is difficult, will be repaired or replaced.

(7) The pack cover will be inspected for attachment of the static line and condition of the static line snap fastener. Static lines

which show cuts, abrasions, broken stitching, or evidence of rust, grease, or acid will be replaced. Defective static line snap fasteners will also be replaced.

(8) Suspension line retaining bands will be replaced if no longer serviceable.

e. The harness will be inspected for damaged of weak spots, especially those that may be produced by rust or battery acid. If the harness webbing shows evidence of wear or deterioration, the harness assembly will be replaced. Inspection will also be made of the stitching and attachment of the harness to the pack. Harness fittings will be examined for breakage, rust, and so forth, special attention being given to the condition of harness snaps and fasteners, the tongues and springs of which sometimes become weakened or bent. If this trouble is found, the snaps or fasteners will be replaced.

SECTION IV

PACKING

1. Repacking Intervals.

a. The main parachute will be inspected and repacked prior to use, if possible, but in no event in excess of fourteen (14) days prior to use.

b. The reserve parachute will be unpacked, inspected, and repacked at least once every thirty (30) days.

c. Regardless of the date when last inspected and packed, parachutes found with evidence of damage to harness or pack will be unpacked and reinspected before being used. Those with stains on the pack will be unpacked and reinspected before being used. Those with stains on the pack from water, oil, acid, and so forth, likely to affect the contents of the pack, will be unpacked and inspected immediately.

2. Folding and Packing.

a. The necessity for carefulness and accuracy in folding a parachute cannot be overemphasized. A misplaced shot bag or hook left in the pack, a knot in one of the suspension lines, or a suspension line dropped while being placed in the suspension line retaining bands are all mistakes easily made, and may result in a parachute failure. Each step in the folding and packing of a parachute should be studiously made, and the operator should give the work his undivided attention.

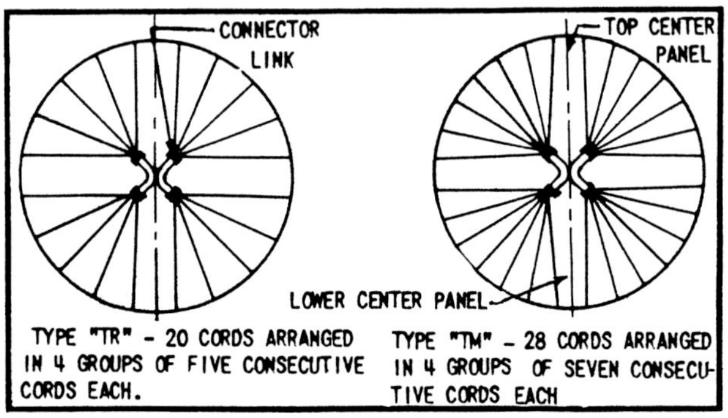

Figure 4 - Arrangement and Numbering of Suspension Lines

b. Referring to figure 4, it will be noted that there is a space through which a vertical center line can be drawn without intersecting any of the lines representing the suspension lines or harness webbing. The points at which this line intersects the circumference locate the centers of two panels that will be referred to as the upper and lower center panels. For convenience in instruction, the suspension lines are shown graphically in figure 4 and are numbered, beginning with the one at the left of the top center panel, counterclockwise, completely around the skirt of the canopy. The suspension lines nearest the center line referred to will be the principal reference lines when the parachute is correctly laid out for folding. (See figures 5 and 6.)

c. The following is the general procedure to be followed upon completion of the operations specified in paragraph 3, section III to accomplish folding and packing the type TM-1 and TR-1 parachutes:

(1) Place the connector links of the harness lift webs over dowel pins in the surface of the packing table to hold the parachute taut. Grasp the suspension lines of the right group in one

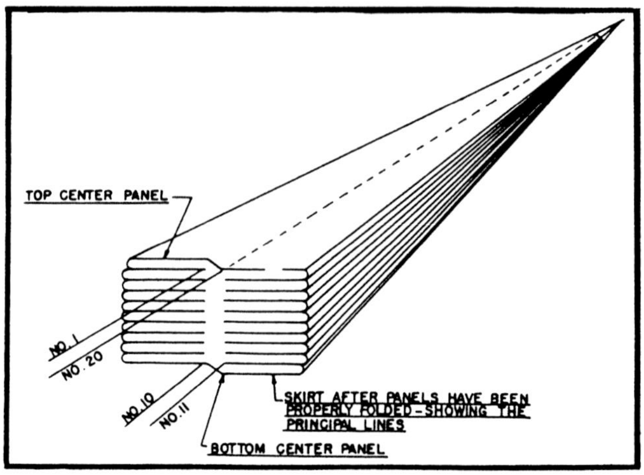

Figure 5 - Folding Diagram - Type "TR" - Reserve

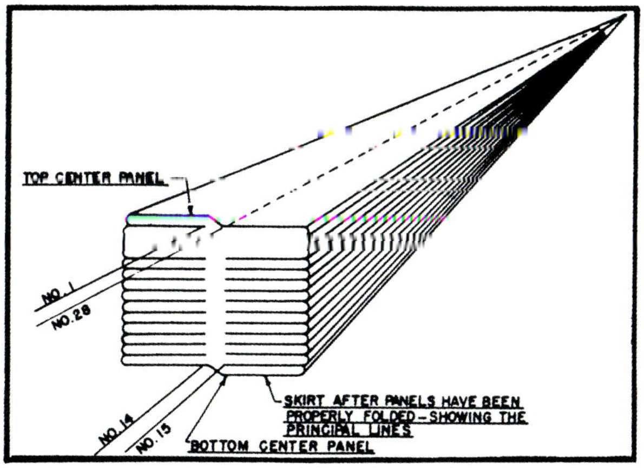

Figure 6 - Folding Diagram - Type "TM" - Main

hand and those of the left group in the other and work toward the skirt of the parachute, thus separating the lines into two groups and at the same time dividing the panels into two groups conforming to the respective suspension lines, as shown in figure 7. Thus arranged, the top center panel (between lines Nos. 28 and 1) should be on the top, and the bottom center panel (between lines Nos. 14 and 15) should be on the bottom, with lines Nos. 1 to 14 inclusive in the left-hand group and Nos. 15 to 28 inclusive in the right-hand group. Next, trace suspension lines Nos. 28 and 1 from the skirt of the canopy back to the harness connector links, ascertaining their correct location. Suspension lines Nos. 14 and 15 should be traced in like manner. The parachute is now in position for folding the panels.

(2) Holding the bottom panel in place on the folding table, place the remaining panels of the right group over to the left, as shown in figure 8, and properly fold this panel (between lines Nos. 14 and 15) by stretching it out smoothly as illustrated. All suspension lines should be kept taut during the folding operation. Fold the remaining panels of the right-hand group in the same manner, placing each panel directly over the other in their correct order, as shown in figure 6. The last panel of the right-hand group to be thus folded will be the one between lines Nos. 27 and 28. After all panels of the right-hand group have been properly folded, place weights on them as shown in figure 9. Next, throw the panels of the left-hand group over those of the right-hand group that have already been folded, so as to expose the bottom panel on the left-hand side, which will be the one between lines Nos. 14 and 13. Fold this panel properly on the table on the left-hand side (see figure 10) and the remaining panels of the left-hand group in their proper order over this one. The last panel of the left hand group to be folded will be the one between lines Nos. 28 and 1, or the top center panel. All panels will then be in their proper positions as shown in figures 6 and 11. By referring to the diagrams of the folded canopy skirt, figures 5 and 6, it will be noticed that no panels cross the center, except the top center panel on the top and bottom center panel on the bottom.

(3) Next, double over each group of panels as shown in figure 12 so that the over-all width of the folded panels is equal to the equivalent width into which the canopy must be placed in the pack. Place weights over these folds in order to hold them in place.

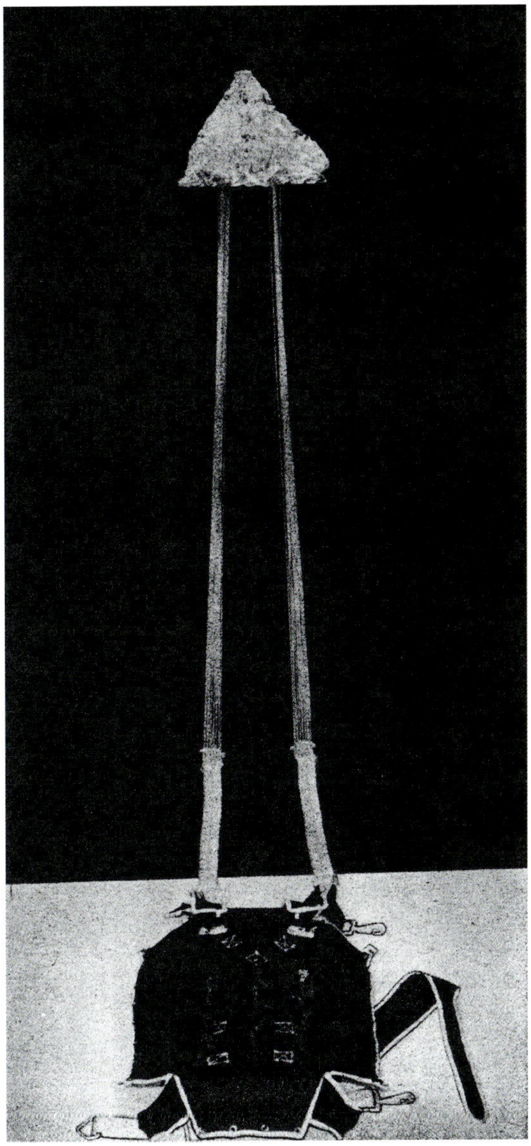

Figure 7 - Canopy, Harness and Pack in Position to Start Packing-Type T-5 Parachute

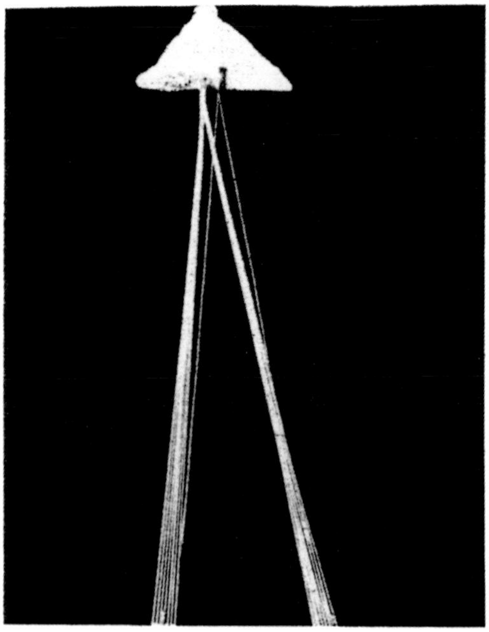

Figure 8 - Proper Position of Panel - Right Hand Group - Type T-5 Parachute

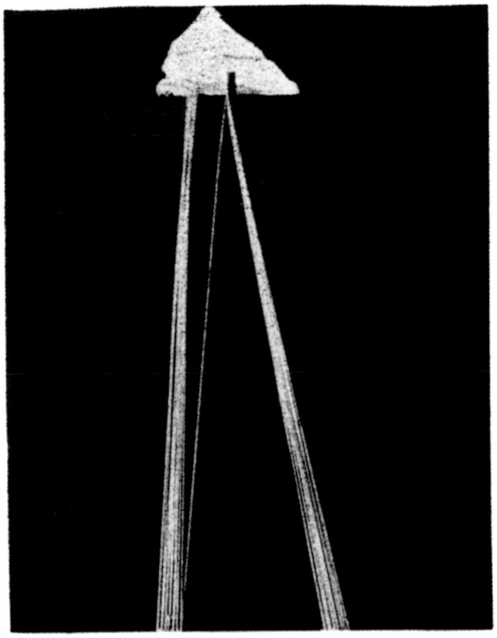

Figure 9 - All Panels of Right Hand Group in Position - Type T-5 Parachute

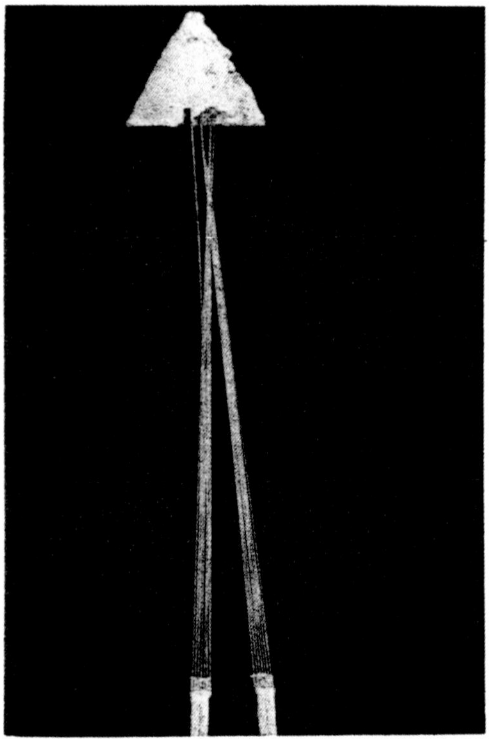

Figure 10 - Proper Position of Bottom Panel Left Hand Group - Type T-5 Parachute

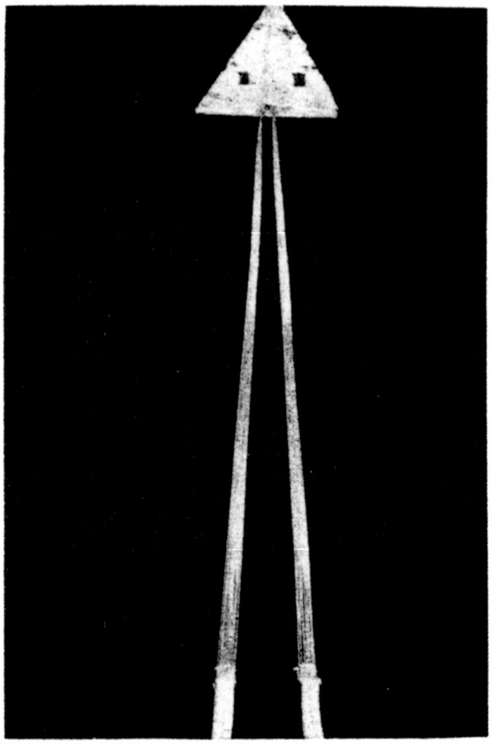

Figure 11 - All Panels in Position For Folding - Type T-5 Parachute

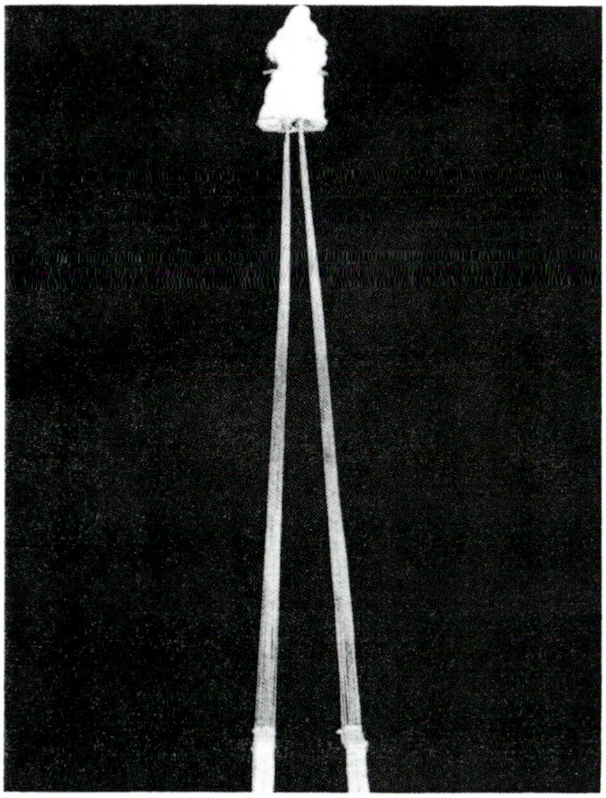

Figure 12 - Panels Folded Over to Proper Width - Type T-5 Parachute

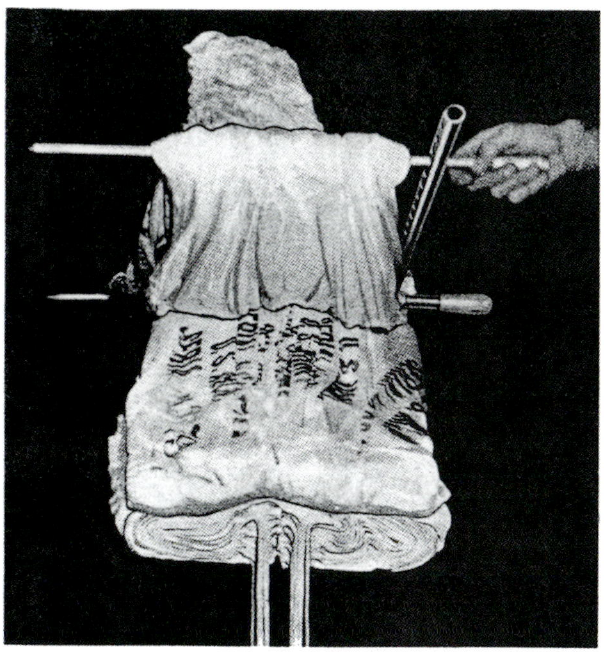

Figure 13 - Folding of Canopy - Type T-5 Parachute

(4) Remove the weights from the folded panels, and release the canopy at the apex. Then fold the canopy in accordion folds starting at the skirt as shown in figure 13, using for this purpose folding tool listed in paragraph 6., section V. Make certain that all weights have been removed.

(5) Place the main left webs so that they extend into the pack, and start inserting the suspension lines as shown in figure 14. The main lift web shall be tacked in two places at the point of entrance to the pack and at the connector link. Tacking will consist of one turn of 5-cord cotton thread at each point.

Figure 14 - Position of Main Lift Webs and Lines in Pack - Type T-5 Parachute

Lift the pack and harness assembly to a position adjacent to the skirt end of the folded panels. Then draw the suspension lines through the rubber loops provided for them, as illustrated in figure 14. Approximately one foot of the suspension lines will be left free between the canopy and pack, this remaining length to be laid loosely across the lines already packed. Make certain that there are no twists or tangles in the suspension lines, as either may cause a failure in the opening of the canopy. The canopy apex is tied to the static line loop with one complete turn of 100-pound silk cord as shown in figure 15. The loop is tied with a doubled half hitch as indicated in figure 1.

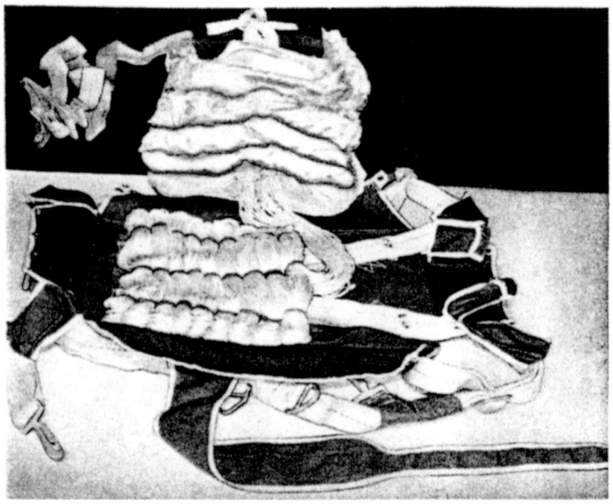

Figure 15 - Lines Completely in Pack Showing Tie Cord Between Static Line Loop and Canopy Apex Cords - Type T-5 Parachute

(6) Place the folded canopy onto the pack, over the folded suspension lines, with the apex uppermost, as shown in figure 16. The pack cover is then laced to the pack flaps with two turns of cotton thread, Spec. AN-V-T-279, passing continuously all around the pack through the grommets located in the cover and the side and end flaps, figure 17. The ends are tied together with a square knot which is safetied by two extra knots.

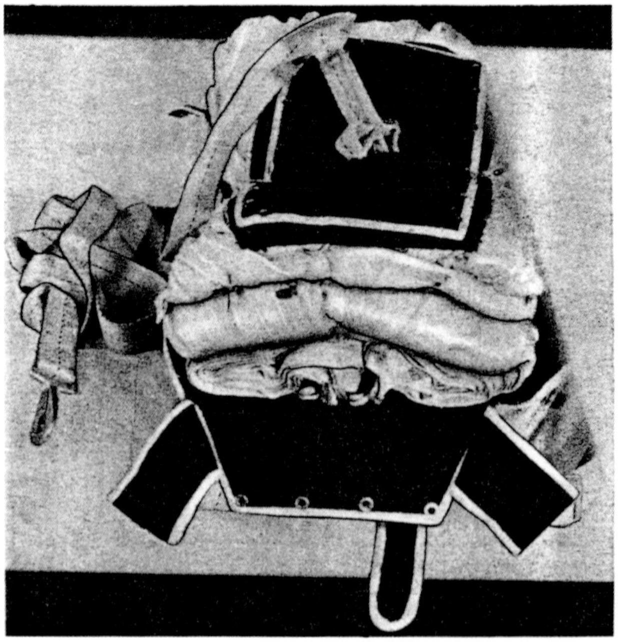

Figure 16 - Canopy and Cover Positioned in Pack - Type T-5 Parachute

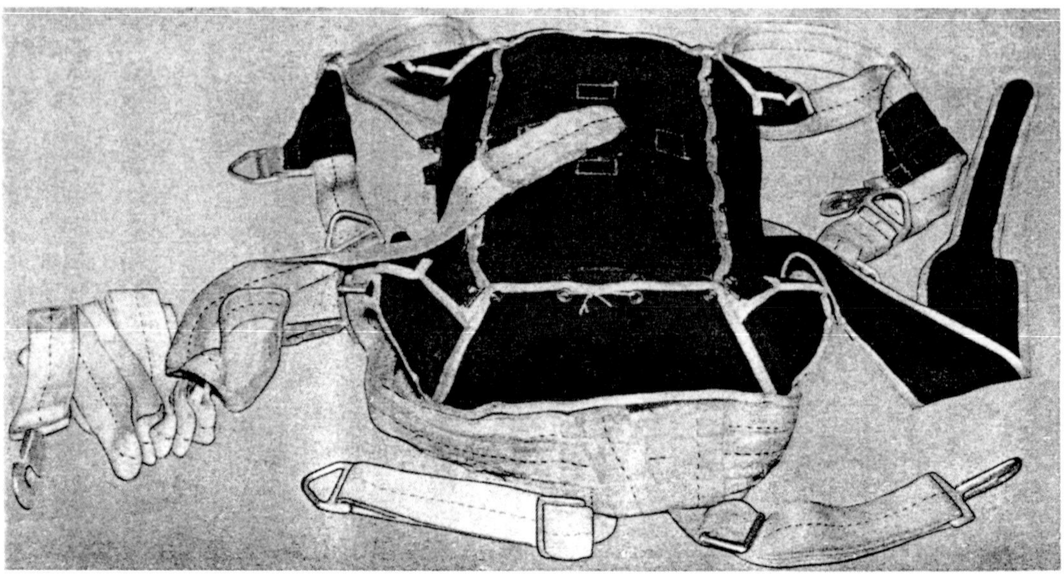

Figure 17 - Pack Load Up and Preliminary Tacking Removed - Type T-5 Parachute

(7) The 15-foot length of slack static line is then stowed into the rubber band retaining bands as indicated in figure 28. When the foregoing operations have been completed, the Parachute Inspection and Drop-Test Card, Air Corps Form No. 58, will be accomplished and inserted in the pocket provided. Instructions covering the use of this card are contained in Air Corps Circular 15-58.

(8) Packing of the reserve parachute will be accomplished separately in a similar manner, reference figures 18 to 27 inclusive.

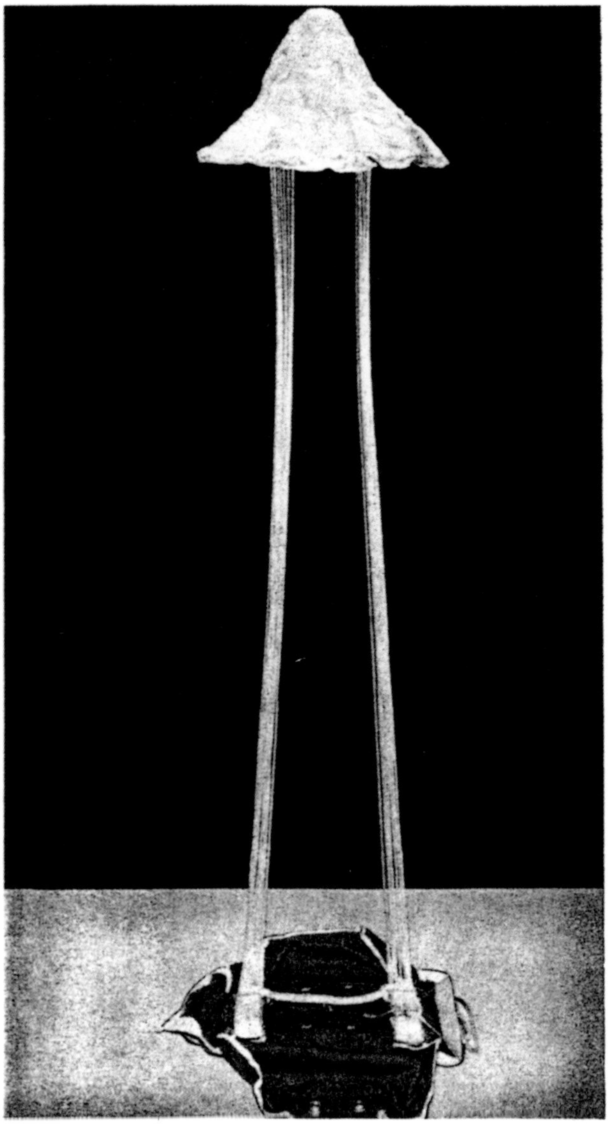

Figure 18 - Reserve Canopy and Pack in Position to Start Packing - Type T-5 Parachute

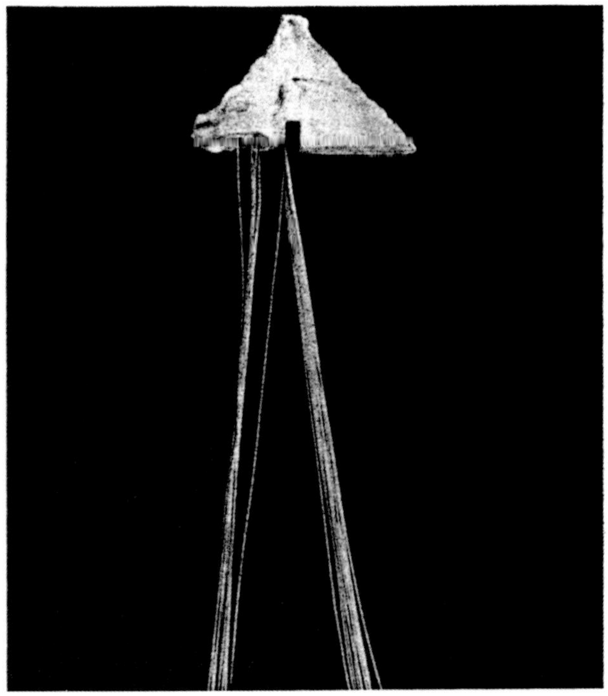

Figure 19 - All Panels of Right-Hand Group in Position (Reserve Pack) - Type T-5 Parachute

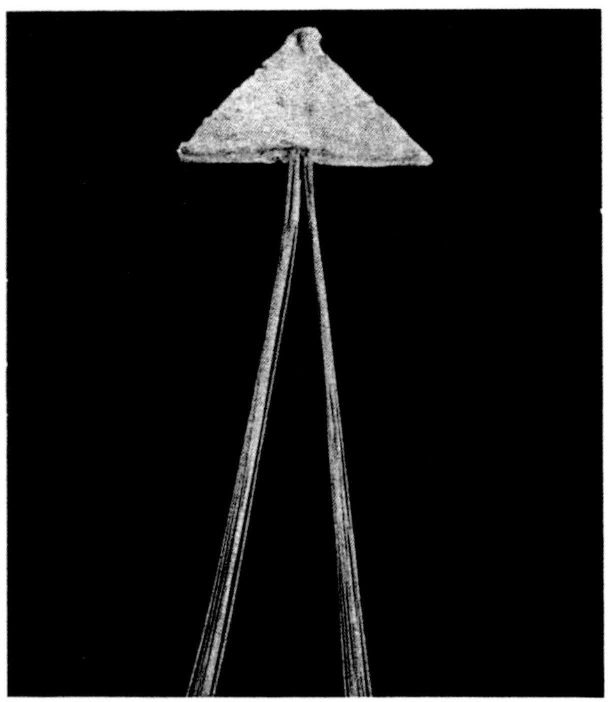

Figure 20 - All Panels in Position for Folding (Reserve Pack) - Type T-5 Parachute

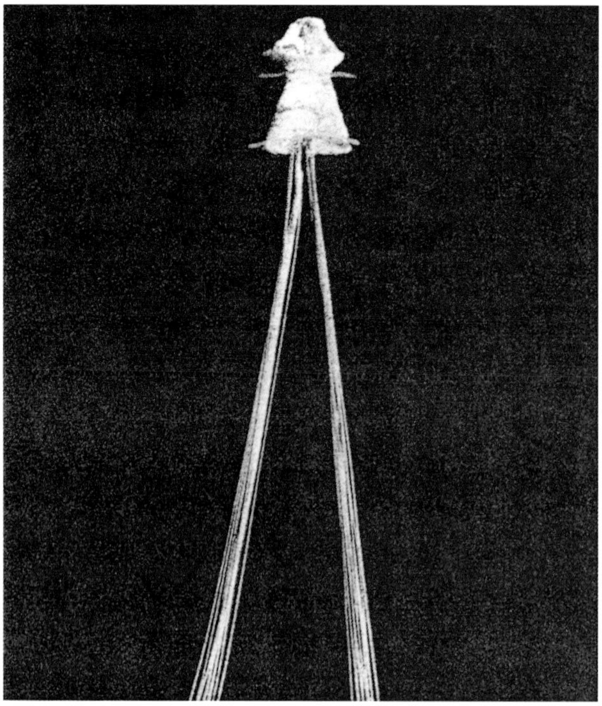

Figure 21 - Panels Folded Over to Proper Pack Width (Reserve Pack) - Type T-5 Parachute

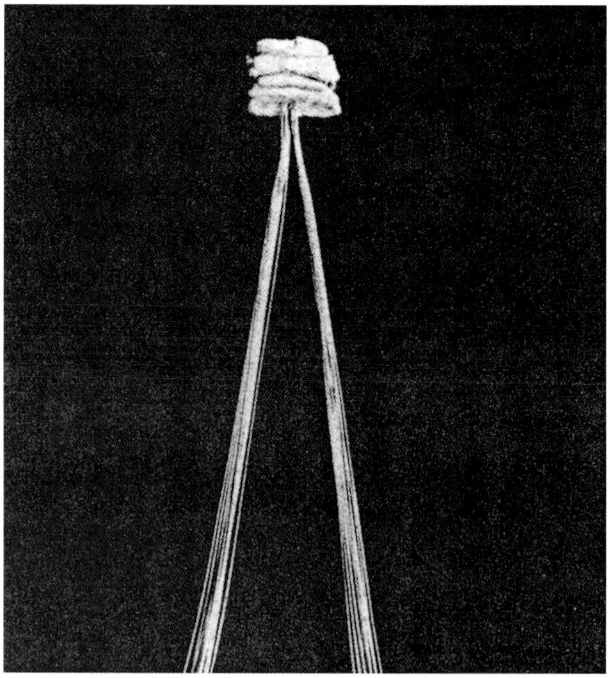

Figure 22 - Canopy Folded in Position for Folding Lines into Pack (Reserve Pack) - Type T-5 Parachute

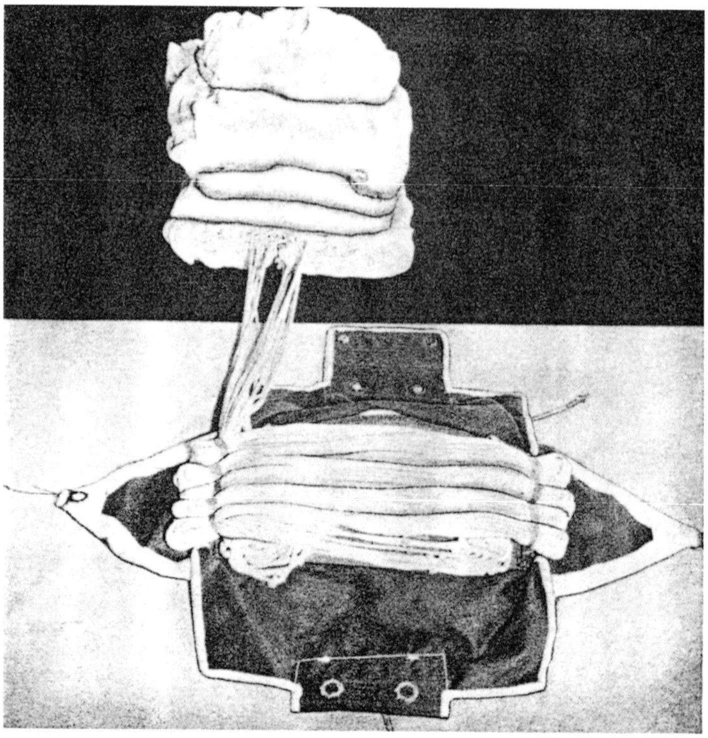

Figure 23 - Suspension Lines Folded into Pack (Reserve Pack) - Type T-5 Parachute

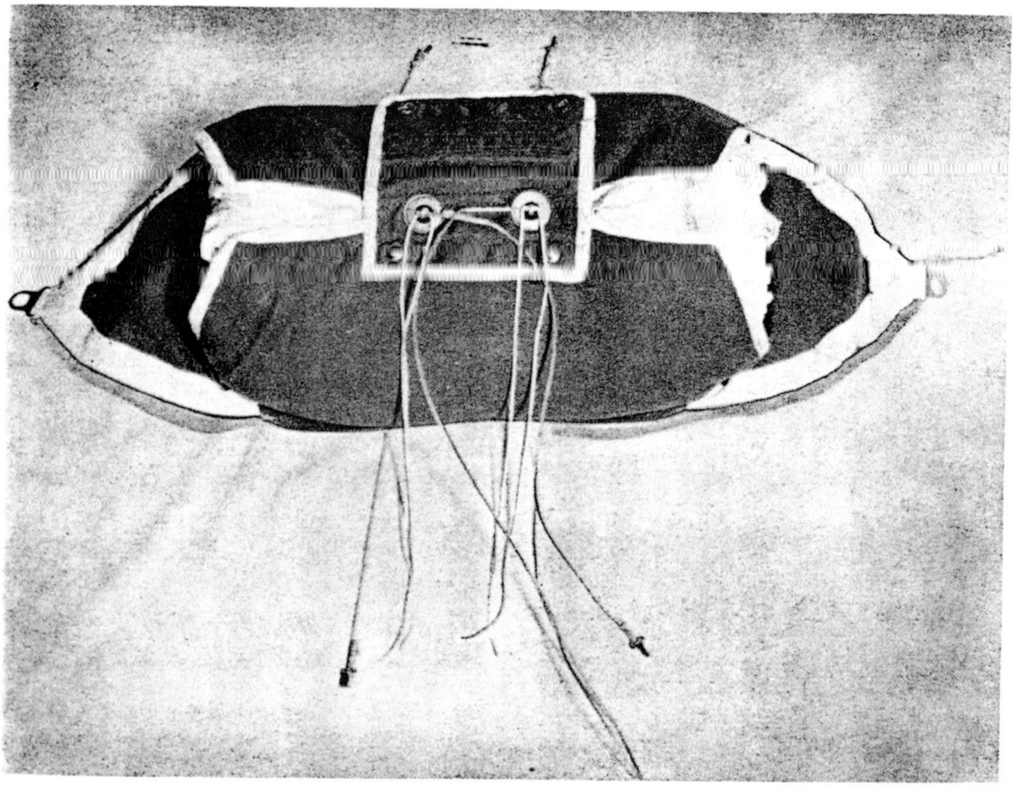

Figure 24 - Side Flaps Folded Over and Pinned With Temporary Rip Cord (Reserve Pack) - Type T-5 Parachute

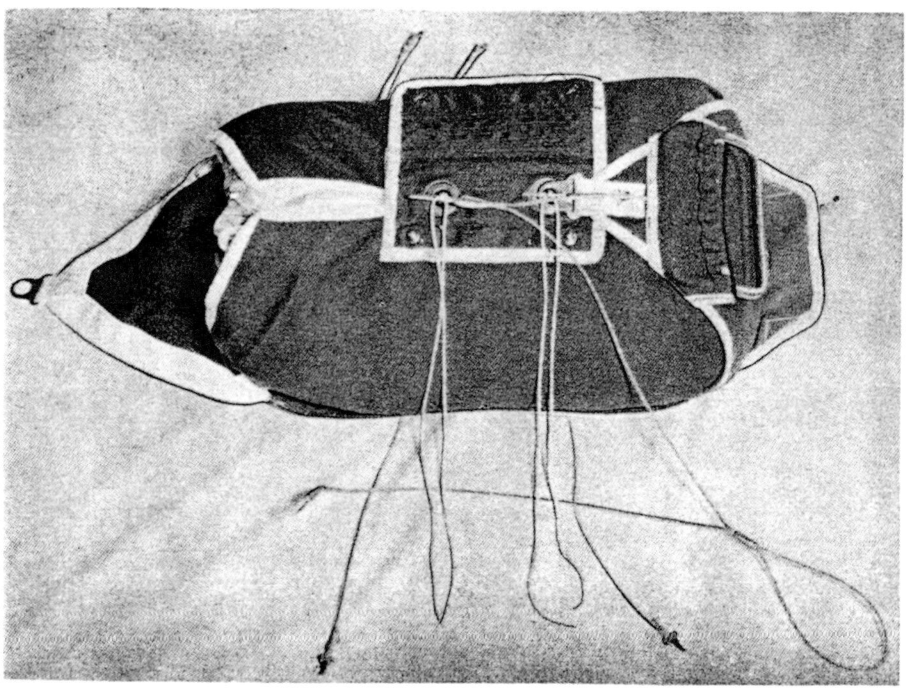

Figure 25 - Folding Over of End Flaps (Reserve Pack) - Type T-5 Parachute

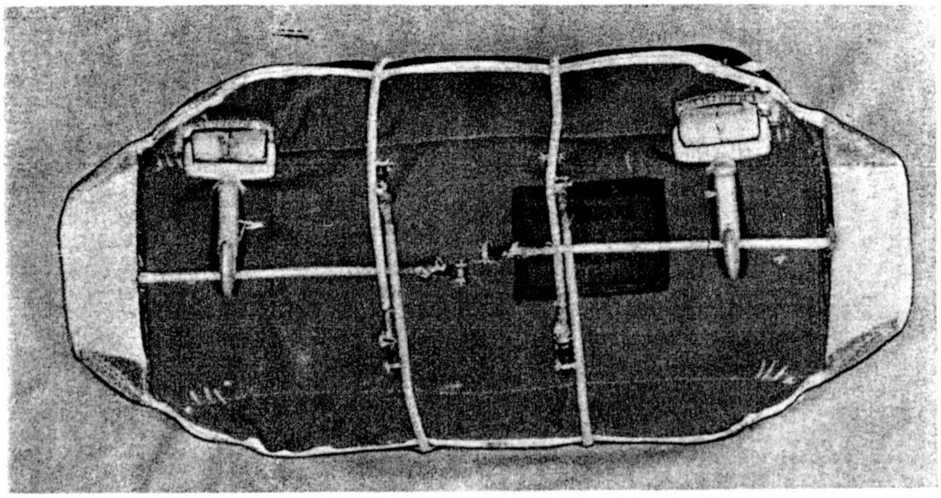

Figure 26 - Reserve Pack Complete (Rear View) - Type T-5 Parachute

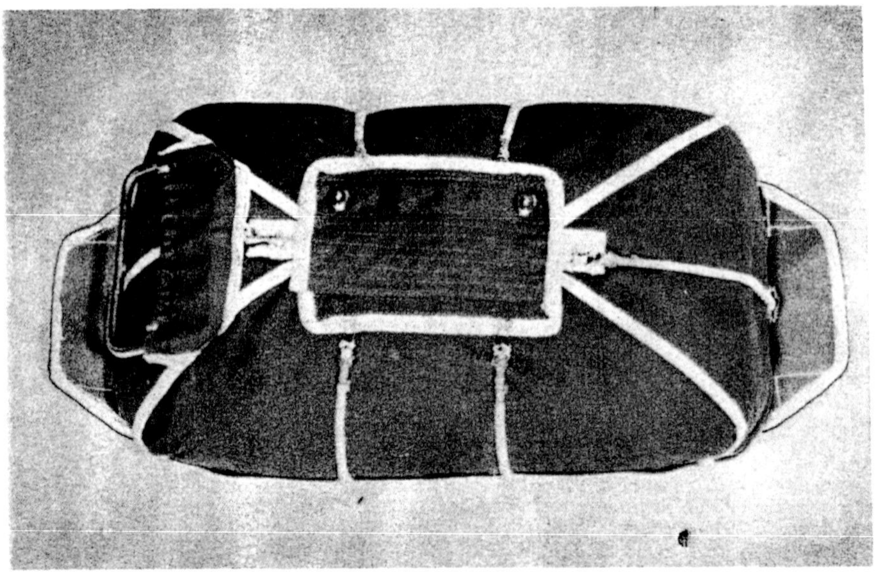

Figure 27 - Reserve Pack Complete (Front View) - Type T-5 Parachute

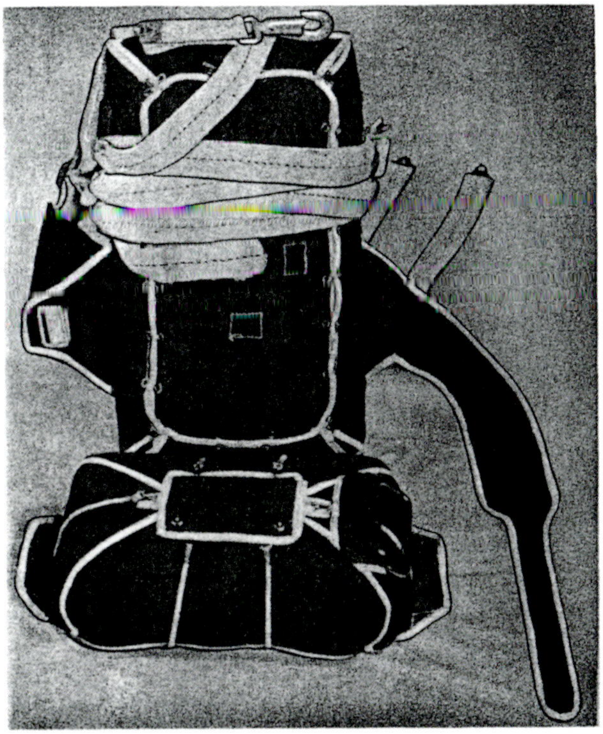

Figure 28 - Type T-5 Parachute - Complete with Static Line Stowed with Retaining Bands

(9) The appearance of the complete assembly with the reserve pack detached as shown in figure 28.

3. Fitting and Adjusting Harness.

a. All parachute harnesses must be fitted to the wearer to provide a snug, comfortable fit, and to prevent loose folds of webbing from catching on projections as the wearer gets in and out of the airplane, also to prevent the lift webs from slipping off his shoulders. Fitting of the harness to the wearer will be accomplished after the parachute has been properly packed. Care will be exercised when issuing organization or operation pool parachutes to ascertain that personnel drawing these parachutes have the harness properly adjusted at the time of issuance.

b. The first operation in fitting the troop parachute harness is to position the adapter at the front of the shoulders as shown in figure 29. The leg straps should be fastened and adjusted so that the leg strap V-ring and snap assemblies are positioned on the upper side of the thigh, reference figure 30. The back straps are adjusted so that the shoulder adapter and the leg strap fittings are maintained in their proper position, reference figure 31. The breast strap is fastened and positioned at a point 12 inches below the chin of the wearer. The reserve pack is then attached by hooking the pack snaps to the harness V-rings and by threading the waist belt from the back pack through the retainer loops provided on the reserve pack, reference figures 32, 33, 34, and 35.

Figure 29 - Harness - Front View - (Unfastened) - Type T-5 Parachute

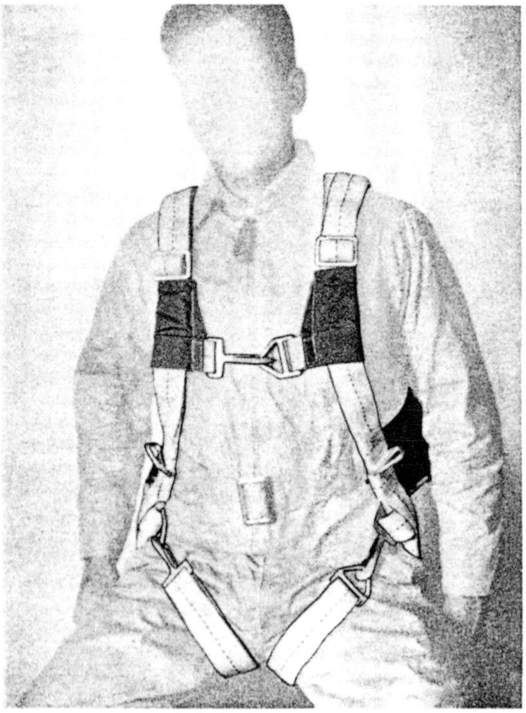

Figure 30 - Harness - Front View - (Fastened) - Type T-5 Parachute

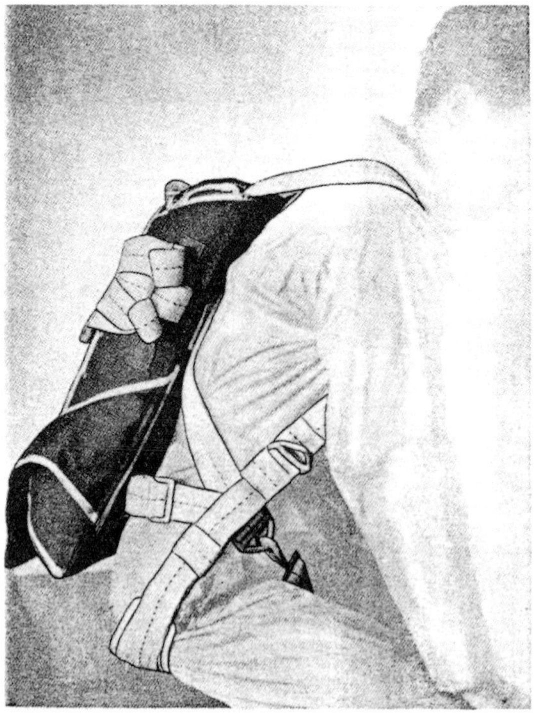

Figure 31 - Harness - Side View - Type T-5 Parachute

Figure 32 - Harness - Showing Attachment of Reserve Pack - Type T-5 Parachute

Figure 33 - Harness - Reserve Pack Attached - Type T-5 Parachute

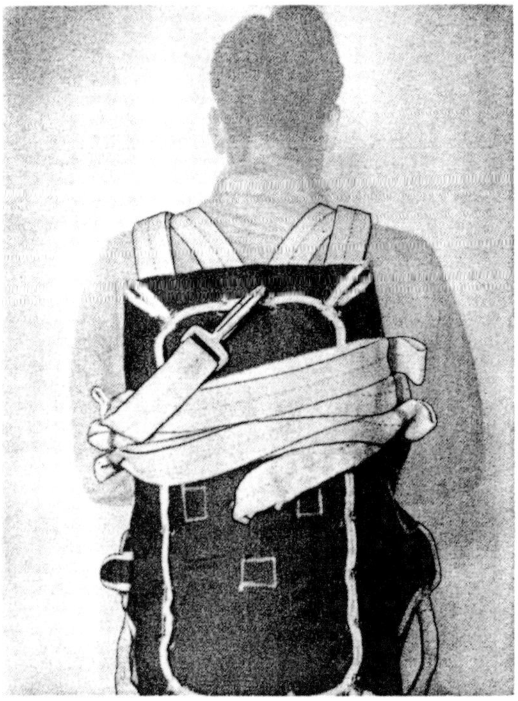

Figure 34 - Type T-5 Parachute (Complete) - Rear View

Figure 35 - Type T-5 Parachute (Complete) - Side View

SECTION V

MAINTENANCE

1. Repair and Overhaul.

a. Repair. - The term "repair" will apply to such operations as replacing pack assemblies, pack-opening rubbers, harness assemblies, vent rubbers, minor repairs to packs as outlined in paragraph 3.e., section III, repair of stitching (except zigzag stitching in canopy), replacement of harness snaps and fittings, and repair of torn places in the canopy, the dimensions of which do not exceed 12 inches in diameter. Such repairs may be accomplished by any activity having qualified personnel (graduates of the parachute rigger's course, Army Air Forces Technical School) and the necessary equipment. Under no circumstances will parachute silk salvaged from parachutes retired from service because of age or deterioration be used in the repair of active parachute canopies.

b. Overhaul. - The term "overhaul" will include in addition to the work outlined in paragraph a. above, all major repair operations necessary to place the parachute assembly in first-class condition. Overhaul will be accomplished by depots only.

(1) Parachutes will be forwarded to the control depot for complete inspection and overhaul, if this is warranted by their condition, the extent of repairs required, or the lack of local repair facilities. Normally, they will be forwarded to the depot when their condition, as disclosed by local inspections (paragraph 3., section III), indicates the need for repairs beyond the scope of paragraph a. above, when they cannot be repaired locally, or whenever doubt exists as to their further serviceability. The date of each depot inspection and overhaul, preceded by the letter "O", will be stenciled on the skirt of the canopy immediately above the data prescribed by Spec. No. AN-A-P-121. Figures and letters 3/4 inch high will be used for this purpose. For example: "O-1-15 39" indicates that overhaul was accomplished January 15, 1939.

(2) Whenever a replacement harness assembly is installed, or a new parachute assembly on which the required harness dates have not been stamped is issued, the date, preceded by the letter "R" will be stenciled on the pack and harness, in the following locations, using letters and figures 1/2 inch high:

Harness - On the right-hand front harness web, at its location on the chest.

For example: "R-2-15-39" will indicate replacement was made February 15, 1939.

Normally, unless visibly defective, all harness assemblies will be replaced only at the expiration of two years service life.

(3) Parachutes will be considered as having a useful life of 100 jumps or four years, whichever is attained first. No repair or overhaul will be accomplished when the cost of the material and labor involved exceeds 50 per cent of the cost of the parachute as shown in the current issue of Stock List, Class 13.

(4) The allowable cost prescribed in paragraph (3) above is the maximum for any one repair, regardless of the number of repairs accomplished. Whenever the cost of any one repair or overhaul exceeds that specified in paragraph (3) above, the parachute can more economically be replaced. Such parachutes, and those multilated in aircraft crashes, may be disposed of locally. Those requiring disposition for any other reason will be forwarded to the control depot for disposal in accordance with Air Corps Circular 65-12.

(5) All parachutes will bear, in addition to the markings required herein, those prescribed in Spec. No. AN-A-P-121.

2. Drop-Testing.

a. The initial drop-testing of parachutes, type T-5, will be accomplished in accordance with the provisions of Spec. No. AN-A-P-121. Parachutes which have had especially hard usage or which are subject to suspicion for any other reason will be drop-tested by the using activity. In the case of parachutes overhauled by depots, the number of overhauled parachutes to be drop-tested will be determined by the Depot Commander or a representative designated by him. Life jumps will be considered as drop-tests.

b. Parachutes may be drop-tested by the using activity by either of two methods, namely, by the use of an airplane equipped with external or internal bomb racks capable of supporting at least 200 pounds, or by the use of a transport type airplane.

(1) Bomb Rack Method. - The standard parachute dummy, part No. 40G7393, will be used. The complete parachute assembly will be fastened to the test dummy by means of the regular harness and the dummy then secured to the bomb rack, head forward, in the same manner as a bomb. For the main parachute, the static line attached to the pack cover will be secured to the bomb rack. For the reserve parachute, the rip cord will be secured to the bomb rack by line assembly, static parachute test, part No. 0161919-1, for external racks, part No. 0161919-2 for internal racks. The parachute will be released from the racks by means of the bomb release mechanism.

(2) Transport Method.

(a) The method outlined in this paragraph is applicable only when a C-33 or C-39 type airplane is used. The complete parachute assembly will be fastened to the standard parachute dummy, part No. 40G7393, by means of the regular harness, and the static line of the main parachute attached to the anchorage cable. For the reserve parachute, the rip cord will be secured to a line assembly, parachute drop-test static, part No. 42B3776, which is attached at the other end to the anchorage cable. The dummy, laid head forward in relation to flight, will be discharged manually from the chute, parachute drop-testing, fabricated and installed in accordance with SAAD drawing No. 41-541.

(b) In the event that the method of drop-testing outlined above cannot be accomplished due to unavailability of equipment, the drop-test may be accomplished by launching the parachute dummy through the door of a transport airplane, simulating the conditions of a live jump, with the parachute static line attached to the anchorage cable within the airplane. Where live jumps are made in lieu of drop-tests, the entries on A.A.F. Form No. 46 shall be the same as for a dummy drop, excepting that data pertinent to the jump, including name of jumper, shall be included under "Remarks".

c. When dropped, the parachute will be observed for promptness of opening and for signs of strain or defects of material. Opening should occur within 3-1/2 seconds. The releasing or discharging of the dummy and parachute will be accomplished at a speed not in excess of 100 miles per hour.

d. After drop-testing, the parachute will be given a thorough inspection and repaired when necessary.

e. The date of the parachute drop-test will be entered on the Parachute Inspection and Drop-Test Card, A.A.F. Form No. 58, and Parachute Record Card, A.A.F. Form No. 46.

3. Cleaning.

a. Parachutes that have come in contact with salt water will, except as noted, be immediately unpacked, and the canopy and suspension lines rinsed in fresh running water for 10 minutes (soft water preferable if available), then suspended and allowed to dry for at least 48 hours or as much longer as is necessary to insure thorough drying. No attempt will be made to wring out the canopy or suspension lines. The harness and pack will be replaced and the removed parts disposed of in accordance with Air

Corps Circular 65-12. Such parachutes will be carefully inspected before repacking and at each subsequent repacking after being placed in service. If any evidence of abnormal deterioration is apparent, the parachute will immediately be shipped to the control depot for disposition. Parachutes that have been submerged in salt water more than 24 hours will be disposed of in accordance with Air Corps Circular 65-12.

b. In the event a canopy becomes badly stained or saturated with oil, grease, or other foreign substances, the canopy will be cleaned in accordance with Spec. 20021 only. This cleaning will be accomplished by service activities if facilities are available, otherwise by local master dry cleaners. In the event the cleaning cannot be accomplished as outlined above, the canopy will be forwarded to the control depot for accomplishment of this work, either locally or commercially. Funds for cleaning purposes, if not available locally, may be obtained under project 85, by requisition. In general, dry cleaning of parachute canopies should be kept to a minimum. Slight discoloration of a canopy from oil or grease may be removed by washing the discolored area with luke warm water and soap, white floating, Spec. P-S-616 (Ivory), followed by a thorough rinsing. Mud should be removed from a canopy by air drying and brushing.

4. Storage and Shipment.

a. Parachutes will be stored in a dry place, protected from the sun's rays, and frequent inspections made for evidence of any deterioration. Except those required for current issue, all parachutes will be stored unpacked, preferably by suspending the canopy from the apex. The silk will not be compressed, or pack-opening rubbers left under tension. No parachute will be left packed at any time or placed in storage when in a damp condition. At any time dampness is evident, the parachute will be thoroughly aired, and dried before repacking or replacing in storage.

b. Parachutes, except as noted in sub-paragraph c. below, will be shipped unpacked with the canopy loosely folded. Within the continental limits of the United States shipments will be made in the flyer's kit bags and the bags enclosed in suitable pasteboard boxes. In the case of shipments to insular possessions, the parachutes will be individually wrapped in wrapping paper and shipped in wooden shipping cases which have been lined with water proof paper. Upon receipt at insular possessions, the parachutes will be promptly inspected. Any instances of parachutes having become contaminated with water will be immediately reported by radio to the Chief, Field Services, Air Service Command, Wright Field, Dayton, Ohio.

c. Parachutes may be shipped packed, if being shipped for immediate use and shipment is made by express or air transport. If shipped packed, the packed parachute will be wrapped in wrapping paper, placed on a flyer's kit bag, and bag enclosed in a suitable pasteboard box. Care should be exercised in the packing for shipment of a packed parachute, that the parachute harness is so folded that no tension will be imposed on the parachute rip cord.

5. Prevention of Mildew.

Fabrics are particularly susceptible to damage by mildew in regions subjected to warm humid climate. Activities in the insular possessions affected by such climatic conditions, and elsewhere, if the presence of mildew is found or suspected, will adhere to the following:

a. All parachutes will be kept as clean as possible, as the propagation of fungi or mold is dependent, to some degree, on such a nutrient as oil, grease, starch, glue, or sizing, in the fabric. The necessary cleaning will be accomplished in accordance with paragraph 3. above.

b. Thoroughly dry in a well-ventilated room at normal temperature before packing. The room should be of sufficient height to permit the canopies to be suspended clear of the floor. Fans will be used to increase the flow of air through the drying room.

c. All packed parachutes not actually being used will be kept in tight lockers or bins, at the bottom of which will be placed suitable containers for dispensing Naphthalene fumes (three pounds of Naphthalene flakes, class B. Spec. R-N-91, class 24, stock No. 8500-624000, to each 10 cubic feet of locker space). In most cases, the fumes from the Naphthalene flakes will penetrate the packed parachute sufficiently to prevent the formation and growth of fungi. Under extreme climatic conditions, approximately 1/8 pound of Naphthalene flakes may be sprinkled over the folded panels of the canopy prior to the final folding operation when the parachute is being repacked.

6. Repairing and Folding Equipment.

A room large enough to accommodate two tables, 20 to 24 feet in length, placed end to end, and with adequate space for suspending canopies for drying is required for inspecting and repacking parachutes. For the accomplishment of this work at stations, the following equipment should be on hand:

- Two tables, 20 to 24 feet long, 3 feet wide
- Bags, shot, 20 inches long, 5 pounds
- Dummy, parachute testing, part No. 40G7393
- Machine, sewing, portable, single stitch, Singer No. 78-1, with combination case and stand
- Machine, sewing Singer No. 112-W-115, double stitch, motor driven
- Awl, round, 4-1/2 inch blade (Osborne No. 1)
- Holder, bench, die, Carr Fastener, No. 194
- Holder, hand tool, Carr Fastener, No. 192
- Needle, sewing, hand No. 1
- Needle, sewing, hand No. 2
- Needle, sewing, hand No. 3
- Needle, sewing, hand No. 4
- Needle, sewing, hand No. 5

Needle, sewing, hand No. 6
Needle, sewing, hand No. 7
Palm, sailmakers' seaming
Press, eyelet-setting, hand power, Carr Fastener, No. M-114
Press, lead seal, hand, parachute inspector's
Punch, cutting, leather, single bow, 9/32 inch
Punch, Carr Fastener socket, hand tool No. 151
Set, grommet, punch and die, No. 5/16 inch
Chuck, Carr Fastener socket, tool No. 1454
Chuck, Carr Fastener socket, tool No. 1456
Die, Carr Fastener clinch plate, tool No. 1455
Die, Carr Fastener washer, tool No. 1457
Tool, parachute folding, chest, lap and seat, type No. 066347-1
Tool, parachute folding, back type, No. 066347-2

SECTION VI

INSTRUCTIONS FOR USING

1. The type T-5 parachute is intended for use in connection with premeditated jumps from airplanes under conditions of controlled flight. This parachute permits landing of personnel in locations where airplane landing is impractical.

2. Although this parachute is capable of landing personnel from a minimum altitude of 200 feet with minimum casualities, such low altitude operation is justified only under special tactical conditions. For training purposes jumps shall be accomplished at not less than 1500 feet altitude and with a maximum ground wind velocity of 10 mph.

3. Detailed technique for leaving an airplane will be governed by the airplane design. In the normal airplane arrangement, jumping will be accomplished from the side door of the fuselage on a twin-engined transport type airplane. These airplanes should be suitably equipped with a static line anchorage cable, non-skid jumping platform at the jumping door, signal lights between pilot and jumper, hand rail under the fuselage, grips at the outside of the door and a launching gear for dropping supplies in the standard containers. The jump will be accomplished in the following manner:

a. The jumper will fasten the static line snap to the anchorage cable in the airplane by removing the snap from its harness position where it is fastened with a break cord and hooked onto the anchorage cable with a quick jerk downward on the left hand. The jumper will then inspect the attachment to see that the static line snap is securely fastened to the anchorage cable.

b. The jumper will then advance to the platform moving the snap along the cable toward the door. Care will be taken to insure that the static line is hooked over the left shoulder behind the jumper and not under the left arm and also that no loop or loose portions of the static line fall in front of the jumper. During this movement the hand will be kept away from the snap tongue at all times.

c. Upon reaching the jumping platform at the door, the jumper will remove his hand from the static line snap and grasp the hand grips at the outside of the door.

d. The pilot will approach the jumping station flying uniformly at constant altitude at a speed of 100 mph and will signal the jumper that the airplane is under control for the jump by means of the green light. Should some unforeseen development occur to make the jump impractical, the pilot will signal with the red light. The flight course will be in a straight line with proper allowance for parachute drift, which shall be calculated by multiplying the windage by the descent period. The descent period in seconds will be determined by dividing the jumping altitude by twenty. Thus, if a jump is made up or down wind with the course directly over the target, the jump should be made short of the target if downwind, and past the target if up wind. When the course must be cross wind the drop shall be either to the left or right of the target as required.

e. When the jumper receives the all clear signal, he will leave the airplane at the proper time as estimated in accordance with paragraph 3. above, or during training operations he will jump at the signal of the jump master. During training jumps, windage will be estimated by placing signal panels on the ground as a target and the use of a parachute dummy drop to determine wind drift. The jump will be accomplished through the door by pulling forward with the arms. The jumper will direct himself laterally out of the airplane. Immediately after leaving the airplane the arms should be brought into the body holding the reserve pack close against the chest and with the right hand over but not grasping the rip cord. The legs should be brought together and the body should fall in a slight crouch in the vertical position with the jumper facing the tail of the airplane.

f. If inflation of the main canopy does not occur within two seconds after jumping, the rip cord on the reserve canopy should be pulled.

g. As soon as the parachute has opened following a jump, the suspension lines should be observed for twists. If twisted, they should be immediately pulled into their proper positions. During the descent the jumper should attempt to face the direction of drift. He should also observe closely his distance from the ground, so as to be prepared to absorb the landing shock with his legs partly flexed. Do not attempt to land in a rigid standing posture or with the back toward the direction of drift.

h. The standard circular type parachute is not maneuverable and only a nominal success is to be anticipated in guiding it. The principal utility of suspension line manipulation is that an increased rate of descent can be realized, thus permitting the jumper to land short of an obstruction. Therefore, all need for such manipulation should be anticipated in sufficient time, and it should not be attempted too close to the ground.

U.S. Marine Corps

The Marine Corps equipped its Parachute Battalions with a different and interesting system. This was the Navy Dual Purpose and was either static line or ripcord operated. It had a row of cones and grommets and a row of grommets through which bungee cord was passed. The left side closed by ripcord pins and the right side by cable pins, under the bungee loops, attached to a static line. The chest mounted reserve was like a standard 24' QAC. This system was manufactured by Pioneer Parachute Co. of Manchester CT.

The Marine Paratroops exited their aircraft in a dive with the left arm wrapped under, while the right arm was over the reserve and the legs held tight together. This posed them in a head down attitude with the canopy billowing out behind.[88] The head down dive aided in extraction of the static line and withdrawal of its closing pin cable. The right hand was positioned to grasp and pull the main ripcord if the jump was to be a freefall. The ripcord only opened the container releasing a pilot chute that hauled the main 28' canopy out. The suspension lines were stowed in S-folds by rubber bands in the bottom of the pack tray conventionally.

The Parachute Marines originally trained at Naval Air Station, Lakehurst, NJ where the Navy had a rigger school. Initially jumps were made from Blimps until transport planes (DC-5) were provided. Later a school was established at New River (Camp LeJeune) NC. Finally, after a school was formed at Camp Gillespie, CA (San Diego), the East coast schools were closed. An additional training facility was created ad hoc in New Caledonia for replacements.

The U.S. Marine Corps can be credited with one combat airborne operation during World War II. In August 1944 an eight man OSS team composed of six Marines, one USAAF officer and a French liaison officer jumped into southern France

Dress tunics of two marines assigned to the OSS. The SF wings indicate assignment to Special Forces HQ.

FIGURE 2-3.

U.S. marine paratrooper's exit technique.

FIGURE 2-4.

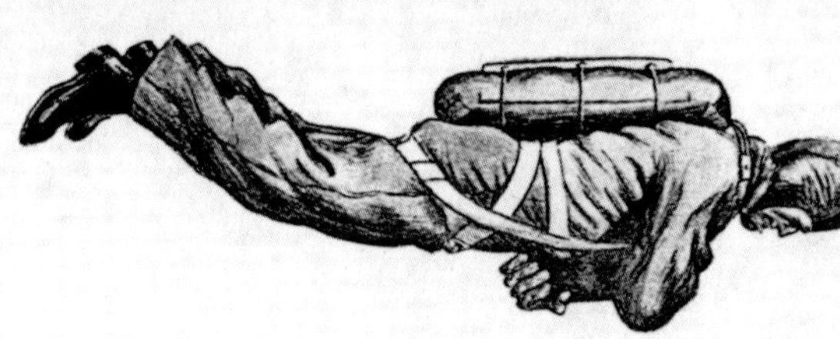

FIGURE 2-5.

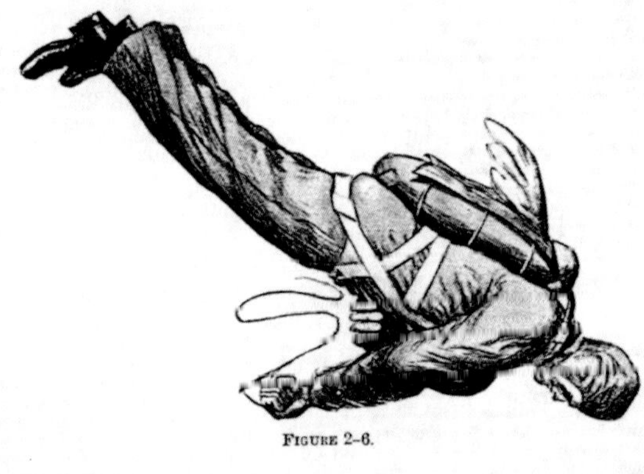

FIGURE 2-6.

prior to and in support of the invasion of the Mediterranean coast. This mission, "Union II"[89] infiltrated using the British X type from B-17s along with a massive cargo drop (899 containers). This mission was led by a Major Peter Ortiz, whose exploits in occupied France as an OSS officer are legendary. This was his second combat infiltration.

The Paratroop Battalions were disbanded in February 1944 and the personnel made up the cadre around which the 5th Marine Division was built.

During the war, the Marine Parachutist badge was the same as the Army's. The gold Naval Parachutist badge was reserved for riggers. The surviving members of the Marine OSS detachment followed the Army practice of embellishing their wings with bronze stars signifying a combat jump.

Navy Dual Purpose

The harness is made of type X cotton webbing with three snap closure. The 24' chest reserve attaches to 3000 lb D-rings. There is no belly band. The pack tray is wire stiffened and is 22" long, 12" wide and about 6" deep. The pack tray fabric is made of a treated or rubberized canvas while the reserve container lacks this weatherproofing. Both containers are dyed a light green.

The static line is a rolled and sewn 13' 10" length of type VIII webbing that is stowed vertically on the ripcord cover flap under two elastic loops. The static line connects by a snap hook to a steel cable with three closing pins. These pins lock the container shut by passing through 3/16" bungee loops routed through the right side grommets. There was a break cord connection between he static line and main canopy with a tensile strength of 100-150 lbs.

The static line opens the main container and releases the pilot chute. This indirect deployment method meant a little more altitude was required for safety. Tests were conducted on New Caledonia to find out how low they could jump and 200' was the limit but unfortunately these produced casualties.

The static line is routed under the bottom of the ripcord protective flap and would be deleted if the jump was to be free fall. Likewise, the ripcord would be unlocked by removing its pins from the cones if the chute was to be static line activated. The static line had snap hooks on both ends. The cable attachment hook is NAF 310336 (Same as 43A21538) and the anchor line snap hook if NAF 310611 (43A21544).

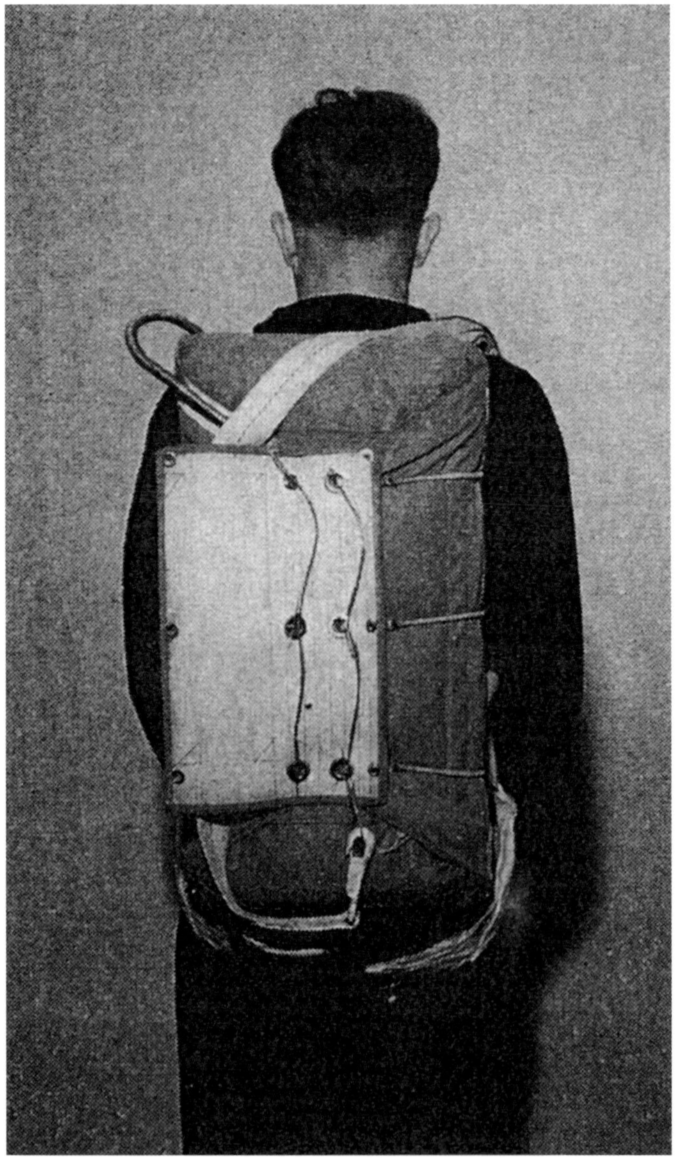

Both canopies appear to be made of undyed silk, which is surprising as the main examined is dated March 1943. A full year after nylon was adopted. Both main and reserve ripcords are clover leaf handled. The main uses a spider type pilot chute. The chest is the same as the 24' QAC, but with the pilot chute removed.

This parachute system was used in combat operations, but not by the U.S. Marine Corps.

Below: Close up of Navy Dual Purpose right side parachute showing details of harness, reserve D-rings, and pack opening elastics. Bottom: Close view of left side of Navy Dual Purpose showing ripcord grip and cable housing. Harness is undyed Type X webbing.

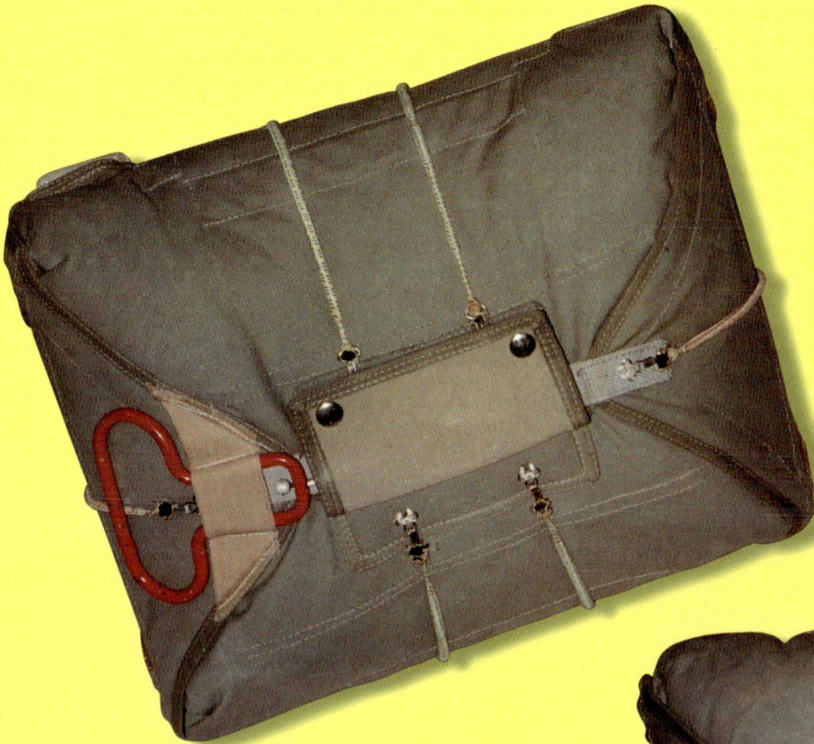

Two views of USN Dual Purpose Reserve. Ripcord pins have been safety wired closed. Static line pin bungee cord is visible in bottom photo.

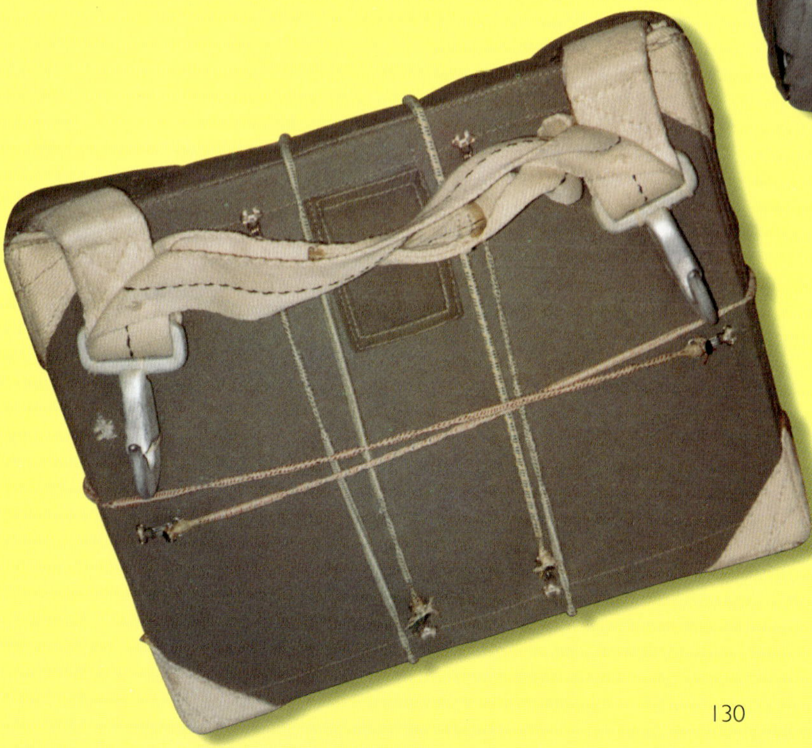

Back side of chest reserve USN DP system showing cross connector strap, pack opening bungees, and packing data card pocket.

This photo shows 82nd troopers wearing the Navy Dual Purpose main affixed to T5 harnesses. Photo taken prior to Sicily invasion, 1943.

Appendices

GLOSSARY

AAF (USAAF) – Army Air Forces 1941-1947, previously Army Air Corps 1921-1941

Adapter – metal rectangular fitting with a crossbar that joins harness webbing.

AGL – Above Ground Level. The altitude that concerns parachutists the most. A variable.

Anchor Line Cable – steel cable inside aircraft to which static lines are attached.

Apex – the very top of the canopy, usually vented.

Assistor Pockets – small fabric pockets fixed to the skirt that fill rapidly and assist in inflation.

"Bang Box" – single point harness quick release assembly.

Becket – a sewn loop to which a temporary connection is made by break thread.

Bias construction – canopy gore fabric warp placed forty-five degrees from the vertical. This was then believed to add strength to the structure.

Canopy – the expanse of fabric that retards the fall of the load. May be circular, square or triangular.

Cones – parachute pack releasing cones. Small chrome plated brass cone shaped hardware that is part of the pack closing arrangement on ripcord operated 'chutes. Plated brass grommets fit over these and a ripcord pin locks the flaps by passing through holes at the top of the cone.

Container – also pack tray if rigid. The outer fabric envelope of the pack.

"Crash and burn" – slang term for a jumpers experience of a hard landing.

Cross connector strap – Also spreader bar. A load bearing capable connection between risers that prevents canopy collapse if one side (hardware) should fail. Found in reserves between the harness snaps.

Data panel – information printed usually on the rear center panel of a canopy that lists maker, serial number, date of manufacture, contract numbers and the like.

Deployment bag – a cloth bag that contains and retains the packed canopy until its suspension lines are fully deployed, also known as the inner bag (British).

Dispatcher – British term for a static jumpmaster.

Drop Zone – the intended landing place for parachutists.

Friction adapter – metal fitting with a sliding crossbar that has a serrated gripping edge for quick fit adjustment.

Glider patch – common term for the round overseas cap patch that combined the parachute and glider symbols. Tradition was for troopers to sew a silver dollar under this insignia, which on occasion, served as a weapon or taxi fare as circumstances dictated.

Gore (formerly panel) – wedge shaped sector of canopy between suspension lines from skirt to apex. Sub divided into panels.

Griswold bag – a zippered, padded case that encloses a disassembled M1 rifle and attaches to the parachute harness under the reserve. Still in service as the M1950 weapons case, though snapped to the reserve D-ring.

"Horseshoe" – A rare, high speed malfunction where part of the canopy is caught on the jumper or his equipment and the rest is, including lines, extended in a great arc thus lending no support to the jumper.

JMPI – Jumpmaster Personnel Inspection.

Jump pay – additional incentive pay for the hazardous duty. The $50.00 per month extra enlisted U.S. troopers received was a hefty sum in the 1940s. Officers got $100.00 per month.

Jump rope – a twenty-five foot long hemp line. Standard U.S. airborne issue, intended for climbing down from trees.

Jump status – condition of service for a soldier or unit whose duties include parachuting.

Jumpmaster – an experienced parachutist who is responsible for the jump operation from the departure airfield to assembly on the drop zone. He may jump with the troops or remain in the aircraft.

Lateral drift – horizontal movement over ground under canopy. Caused by wind and relieved by jumper pulling a slip into the wind.

"Leg" – pejorative term for all non-jumpers. From 'straight-leg' referring to a lack of jump boots and the uninterrupted trouser leg.

Leg bag – A British innovation for combat equipment accompanying the jumper. Attached to parachute harness by a 20' rope and hand held by the jumper through the exit door. It resembles a short duffel bag with lace up closing.

Lower Lateral Band – canopy skirt.

"Mae West" – (also line over, blown periphery, partial inversion) a low speed malfunction in which the skirt is pushed by the slipstream through the lines on the other side, creating a lobe that inflates at the expense of the canopy surface area. Often asymmetrical, but when equal lobes are produced, it will appear to be a huge brassiere, hence the name (from a rather well endowed actress of the 1930s and 1940s).

MSL – Mean Sea Level. The theoretical plane from which aviators reckon their altitude. A constant.

Musette bag – a standard U.S. item of field equipment. Attached by troopers to harness under the reserve.

Oscillation – pendulum like swinging of suspended load under a developed canopy. Caused by trapped air venting under skirt. Relieved by jumper pulling a slip. Unchecked, it usually gets worse.

Outer Bag – container.

Pack opening elastics – rubber bungee cords that apply spring tension to a container flap and pull it away from the enclosed canopy upon activation.

Pack opening spring bands – same as above except incorporate metal coil springs. Adopted in 1948 from a German design, to ensure function in extreme low temperatures.

Panel – individual cloth unit of canopy, sub section of gore between radial seams.

Pathfinders – Paratroops who precede the main body and mark the DZ with panels, lights and homing beacons. An additional skill qualification.

Permeability – rate at which air passes through a fabric.

Pilot chute – a small parachute that is deployed in advance of the main. It serves as an anchor and pulls the main canopy out of the container. May or may not be spring actuated.

PLF – Parachute Landing Fall. The every best way to absorb the impact of landing. Devised by Flt.Lt. John C. Kilkenny RAF in late 1941 – early 1942. It is the placement of the fleshy body areas in sequence on the ground: balls of the feet, side of the calf, thigh, buttocks, side of back, with the elbows tucked in close and head lowered. Expertly done, it produces a roll that rebounds the jumper back to his feet where he can run around his canopy to collapse it to avoid wind drag.

Porosity – a ratio of voids to a volume or area. Often mistaken for permeability.

Pucker bands – (German 'taschengurts') a short tape sewn at the skirt from gore to gore that draws them together to pucker the gores in order to speed inflation. Added to T7 reserve making T7A.

Pucker vent – elastic O-ring at the apex that released high pressure air at canopy opening, then closed to trap air when the canopy was fully developed.

QAC – Quick Attachable Chest. An aircrew escape system in which the airman wears a harness and clips on the appropriate chest type, then bails out.

Quick release assembly – "bang box", "dial of death" – a single point release device that connects the leg and chest straps and was considered an advancement because of its faster egress

Ripcord – a steel cable to which pins are affixed that closes a parachute pack.

Ripstop – a postwar type of weave designed to resist tears. Identified by its continuous box pattern.

Risers – extensions of the harness to which suspension lines are attached. Called "main lift webs" by the British.

"Roman Candle" – British term for a streamer.

Serving – a tight binding of thread or cord around a rope holding it together.

Skirt – also lower lateral band, the reinforced bottom hem of the canopy.

Squidding – A short phase in canopy development during line stretch where the top of the canopy is filling but the lower half is still closed. So called because it vaguely momentarily resembles the sea creature.

Static line – a line connecting the parachute pack to a point, usually a cable, in the aircraft that automatically begins the deployment of the parachute as the load falls away.

Stick – a line of parachutists aloft or in the plane.

Stow loops – (hesitator loops) cloth or web loops that hold the suspension lines in a neat arrangement on a deployment bag or in a pack tray and pay the lines out under some tension during deployment.

Streamer – (British: "Roman candle" from its flickering flame-like appearance) a partial, high speed malfunction in which the canopy is stretched out but does not inflate. One of the very worst possibilities, but quite rare.

Suspension lines – also shroud or rigging lines, tensile rated cords that connect the canopy to the load.

Total – a malfunction where nothing comes out. Caused by cut static line, broken snap hook, anchor cable defect or failure of jumper to hook up. This high speed event has a good aspect – there is nothing above the jumper to entangle his reserve or capture its pilot chute. An extremely rare event.

Warp and woof, -weft, -fill – Three ways of saying the same thing. Warp are the long threads in a weave and run parallel to the selvage edge. Weft are the short threads in the weave.

Wind channel – the central vertical core of a pleated canopy where the radial seams (and suspension lines) are loosely bundled together.

Wings oval – a decorative cloth background for the parachutist badge in Regimental colors that serves as a jump status designator.

FAKES

There is such a thing as a fake wartime parachute. This author has seen two such and both were facsimiles of the U.S. T5. The first example proved on close examination to be a reworked A2. The harness being a QAC would appear to be right, but the pack tray was wrong, looking suspiciously like a back crew escape with a jerry-rigged static line.

The second was sold to an associate as being a complete correct T5. At a glance, one could see it was not complete, as the pack tray was not filled out. The packed T5 is a full, tight bundle. The data panel on the pack tray appeared to have been lifted from an authentic T5 and poorly sewn onto the notional one. The pack tray cover was a obvious reproduction (well done) with a new static line but with the modern T10 slide snap hook. The cover was not correctly laced to the tray.

All doubt vanished when the cover was removed to reveal swatches of mutilated T7 canopy fabric, with nylon OG taffeta and a de-mil'd canvas shelter half. It was there as filler with the camo cloth on top so any casual inspection would lead one to believe that there was a canopy within.

This writer also found a 1945 T7 reserve container represented as complete, but pulling back a corner flap revealed the bright red cotton of part of a cargo canopy. The seller just did not know what was in there and most owners are afraid to open these 'chutes up fearing they won't be able to put them back right.

There is nothing wrong with reproducing anything as long as the purchaser is aware that the item is not an original. The most difficult part to find for a T5 is the tray cover-static line assembly. These have been reproduced to specification by at least one expert fabricator. They are not sold as originals, so no fraud is committed. Their pristine condition betrays their newness.

Collectors and exhibitors will almost assuredly have to assemble a T5 or T7 from parts. Nothing wrong with that as that was how they were manufactured. A harness can be made by capable hands from 'scratch'.

No compromise with authenticity as many harnesses were made that way by the Parachute Maintenance Companies. The canopy can be substituted for with any 28' flat circular that still has its lines (the Air Force C8 in solid white would be ideal). The T10 canopy would not do as it is too big.

The only hard thing to make is the pack tray. This really is sine qua non and it is highly doubtful if anyone will put those back into production. Luckily there are still a good number of T7 pack trays about and there are still numbers of T5 reserve containers on the market.

These reserves can be filled with surplus T10R canopies. The standard T10R ripcord handle will fit as they are the same. The snap hardware is still available from some of the parachute supply houses and there is little demand for it.

Be aware that early T5 reserves had a 22' canopy. Look for stamping on the outside '24 ft. diam.'

So that you don't destroy the container by trying to force a 24' canopy into a space intended for a 22' one.

Additional clues: the ripcord pocket is for a trapezoidal handle, the drawing number is 42K4960-1 (also 065311), early date 1941 or 1942 and a 24' canopy is an extremely tight fit.

The fifty lb. test cotton lacing cord does not seem to be available anymore, but a reasonable substitute can be found in a local hardware store. This material is like string and not at all like the eighty lb. test tape used on the current T10 series of parachutes.

The phony T5. The harness appears to be an original in poor condition. The pack tray is most likely a T7 to which a T5 data panel was inexpertly sewn. The pack tray cover is most definitely a reproduction and the canopy is simulated by an inadequate amount of scrap fabric.

The static line looks too new.

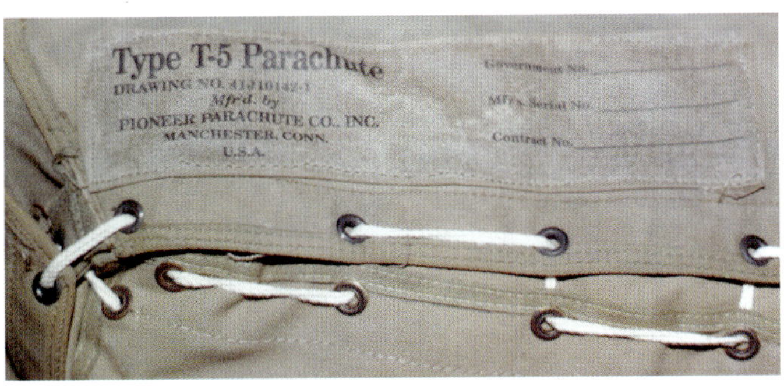

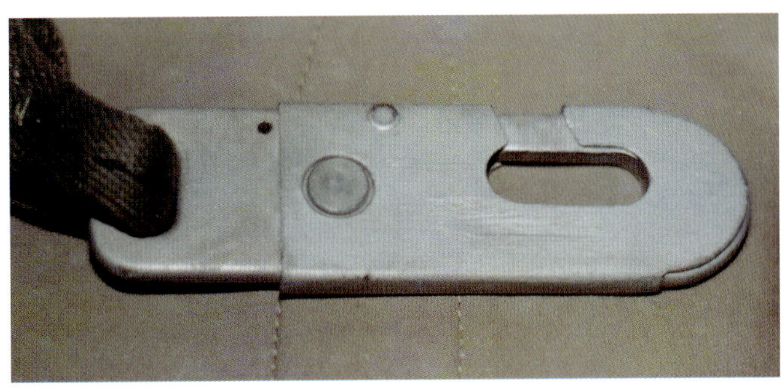

Notice where the MS 70120 number was scraped away on the gate sleeve.The tell-tale safety wire hole could not be removed.These were initially (1947) at the top of the hook.

FATE

People ask "What happened to them all?" A fair question since many thousands of troop parachutes were produced. Many were consumed during combat operations and not salvaged. Many went to armies in the "Third World" where they were used up. A few days exposure to the elements, especially in the tropics, is sufficient to weaken the fabric to uselessness. Locals prized them for the cloth.

The equipment is not valued for what it is among the masses and government minions are loathe to part with them to mere citizens in usable condition. So the lines are chopped, panel slashed, webbing cut before being sold for scrap.

The civilian populace is no better; many a canopy has been desecrated as a poor and very short lived car cover or hung in a dormitory as trendy decoration. Incredibly even some surplus dealers have been observed chopping old T5 harnesses to extract the hardware (!) What they mindlessly destroyed is now worth thousands of dollars and can be duplicated only at a great deal of trouble, if at all.

The wanton destruction is symptomatic of an appalling ignorance of history, a disregard for posterity and an indifference to one's heritage. The U.S. government has an especially wretched record as custodian of artifacts and has displayed a destructive enthusiasm approaching the vandals of antiquity. To wit: they made corduroy roads of Confederate muskets, they burned every Fokker D VII at hand, they scrapped every B-32 Dominator, they chopped up almost every .30 Pedersen Device and torch cut tons of small arms after Desert Storm and each time as if in a rage of revenge.

Those who really care about history must guard its tangible relics with a jealous care greater than the zeal of those who would cheerfully obliterate them. It is our heritage, and our predecessors deserve no less.

WARTIME MANUFACTURERS
of troop parachutes or components include:

Aerial Machine & Tool Co. – Queens, Long Island, NY. Produced hardware only.
American Lady Corset Co. – Chicago, IL
Atlantic Rayon/Atlantic Parachute – Lowell, MA
Cheney Bros. – St.Louis, MO
Cole of California – Sacramento, CA
Eagle Parachute – Lancaster, PA
Sigmund Eisner – Red Bank, NJ
Fashion Frocks – Cincinnati, OH. The largest maker of pack trays.
General Textiles Mills – New York City, NY
Hayes Manufacturing. Co. – Grand Rapids, MI Hightstown Rug Co. – Hightstown, NJ
Irving Air Chute – Buffalo, NY. Lexington, KY.
Lady Mac Corset Co. – Buffalo, NY
I.G. Menihan – Rochester, NY
I. Miller – Long Island, NY. Produced canopies.
National Automotive Fibers – Detroit, MI
Otterbein – Kensington, CT. Produced hardware. Only two companies made parachute hardware.
Pioneer Parachute Co. – Manchester, CT
Reliance Manufacturing Co. – Chicago, IL. The largest single T5 maker with three plants in two states (Indiana, Mississippi).
Security Parachute Co. – Oakland, CA
Simmons Co. – Elizabeth, NJ
Standard Parachute – San Diego, CA
M. Steinthal – New York City, NY
John B. Stetson Co. – Philadelphia, PA
Switlik Parachute Co. – Trenton, NJ
Triangle Parachute – Cincinnati, OH
Vanity Fair Mills – Reading, PA
Victory Parachute Co. – Scranton, PA

It may be impossible to determine who made what, where, when and in what quantity now as some of the wartime firms are many years out of business or combined with others. Sadly, some companies are known to have thrown out their archives after they were absorbed into others.

Sources

AAF manual 55-0-1 " Reference Manual for Personal Equipment".

"Airborne Equipment and Airborne Defenses of the Soviet Union" Dept of the Army, Wash. DC July 1955

Andrews, John C. "Airborne Album vol. I" Phillips Publications, Williamstown NJ 1982.

Anon. "By Air to Battle" HMSO London 1945.

Bando, Mark "101st Airborne: From Holland to Hitler's Eagle's Nest" Motorbooks Osceola WI 1995

Blandford, Edmund "Green Devils – Red Devils" Leo Cooper London 199

Bradley, Francis "Paratrooper" Stackpole, Harrisburg PA 1967

Bragg, R.J. & Turner "Parachute Badges and Insignia of the World" Blandford, Poole, Dorset 1979

Bouchery, Jean "Les Paras US de Normandie" Militaria magazine July/Aug 1987

Brown, W.D. "Parachutes" Sir Isaac Pitman & Sons London 1951

Burhans, Robert "The First Special Service Force" Infantry Journal Press Wash. DC 1947

Crookenden, Napier "Airborne at War" Chas. Scribner's Sons, New York 1978

Crookended, Napier "Drop Zone Normandy" Chas. Scribner's Sons New York 1976.

Davies, Howard P. "British Parachute Forces 1940-45" Arco Pub. New York 1974

"Development of Technique of Dropping Troops and Military Supplies by Parachute 1939 – 1945"

Airborne Forces Experimental Establishment Report No. 28, 28 Feb 1949

"Development of the Parachute and Airborne Activities in the USSR" Col. Charles J. Timmes US Army G 2, June 1952

De Trez, Michel "American Warriors" D-Day Publishing, Wezembeek-Oppem, Belgium 1994

Devlin, Gerard "Paratrooper!" St. Martin's Press New York 1979

Ellis, John T Lt Col "The Airborne Command and Center" Study No. 25 Historical Section

"Establishment and Operation of a Parachute School in a Theater of Operations" The Parachute School

Ft Benning GA, 26 March 1943

Gabel, Kurt "The Making of a Paratrooper" Lawrence KS 1990

Glantz, David "Soviet Airborne Experience" Leavenworth KS 1984

Greenwood, Jim "Parachuting for Sport" General Publishing Co. New York 1962

Gregory, Barry "British Airborne Troops" Doubleday & Co., Garden City NY 1974

Gregory, Barry "Airborne Warfare 1918-1945" Phoebus, London 1979

Hall, Roger "You're Stepping on My Cloak and Dagger" WW Norton, New York 1957

"Handbook of Instructions for Parachutes – Troop Type TO 13-5-7" Wright Field, Ohio Sept 1942

Haney, Ken "US Marine Corps Paratroopers 1940 – 1945" Eagle,Globe Anchor Pub. Jackson TN

Harclerode, Peter "Go to It!" Bloomsbury, London 1990

Hickey, Michael "Out of the Sky" Charles Scribner's Sons, New York 1979

Horan, Michael "Parachuting Folklore" Parachuting Resources, Richmond IN 1980

Howard, Gary "America's Finest" Stackpole Books, Mechanicsburg PA 1994

Hoyt, Edwin "Airborne" Stein and Day, New York 1979

"Illustrated Catalog Class 13 Supply (clothing, parachutes, equipment and supplies) USAAF 30 Sept 1943

Jonca, Adam "Polscy Spadochroniarze" Wydawnictwo Bellona, Warsaw 1994

LIFE magazine 19 June 1944

Low, Archibald "Parachutes in War and Peace" London 1942

Mackersey, Ian "Into the Silk" Robert Hale Ltd, London, 1956

McCaul, Ed "82nd Airborne Paratrooper at Normandy" Military History magazine June 1997

Mink, O.J. "Meet the Parachute" Reliance Mfg Co. Chicago IL 1944

Nalty, Bernard "D-Day Operation Overlord" Smithmark, New York 1993

Newnham, Maurice "Prelude to Glory" Sampson, Low, Marston Co. London 1947

Otway, TBH "Airborne Forces" War Office, London 1951

"Parachute Instruction Manual. Marine Parachute Troops" Camp Gillespie San Diego CA 1943

"Parachute Rigger 3c & 2c" Navy Training Course, Bureau of Navy Personnel GPO 1943

"Parachute Training Manual" Air Ministry, London 1944

"Parachute School USA" Albert Love Ent. Atlanta, GA Sept 1943

"Parachute Training" souvenir item Ft Benning GA 1944

"Parachute Training" souvenir item , M.D. Boland, Tacoma WA 1943

Poynter, Dan "The Parachute Manual" Parachuting Pub.s, Santa Barbara CA 1977

RAF Airborne Forces Manual 1942 Arms & Armour Press, London 1979

Ste. Croix de, Philip "Airborne Operations" Salamander Books, London 1978

Smith, Floyd "Parachutes" Manchester, CT 1942

"Supply by Sky: the Quartermaster Airborne Development 1950-53 Office of the Quartermaster General, Dept of Army 1954.

"Students Text" The Parachute School, Airborne Command Ft Benning GA 1943

Sweeting C.G. " Combat Flying Equipment US Army Aviators Equipment 1917 – 1945" Smithsonian Institute Press, Wash. DC 1989

"Tactics and Techniques of Air-borne Troops" FM 31-30 War Dept 20 May 1942

TM 1-440 "Parachutes and Aircraft Clothing" War Dept. June 1945
TM 10-1670-201-25 "Maintenance of Parachutes and Other Airdrop Equipment, General " Dept. of Army, Dec 1965
TM 71-220 "Technical Training of Parachutists" War Dept. Sept 1946
TM 71-220 "Technical Training of Parachutists" Dept of Army Oct 1948
Thompson, Leroy "US Airborne in Action" Squadron/Signal Pub.s Carrollton TX 1992
"Training of the Parachutist" DOSAAF Printing House, Moscow USSR 1954
"Try Nylon for Parachutes" Science News Letter 15 July 19 page 40
Tugwell, Maurice "Airborne to Battle" William Kimber, London 1971
Weeks, John "Assault from the Sky" G.P.Putnam's Sons New York 1978
Weeks, John "The Airborne Soldier" Sterling Publishing New York 1982
Weeks, John "Airborne Equipment" Hippocrene Books New York 1976
Wintersteen, H.G. "Practical Flying" International Textbook Co. Scranton PA 1943
Wright, Ian B. "Bale Out!" GMS Enterprises Peterborough England 1991
"X-type Assembly Mark 4" (packing instructions) Ministry of Defense March 1960
Zaloga, Steven "Inside the Blue Berets" Presidio Press Novato CA 1995
"Specialized Supplies Dropping Apparatus" S.D. 0373 Air Ministry October 1942

Notes

[1] Ian Wright "Bale Out!" GMS Ent., Peterborough England 1991, page 20.

[2] FM 100-2-2 "The Soviet Army Specialized Warfare & Rear Area Support" 16 July 1984

[3] Zim, Herbert "Parachutes" Harcourt Brace, New York 1942, page 134.

[4] FM 100-2-2 op cit.

[5] David Glantz "Soviet Airborne Experience" Command & General Staff College,Ft Leavenworth KS 1984 , page 4.

[6] Steven Zaloga "Inside the Blue Berets" Presidio Press Novato CA 1995, page 7.

[7]5"Training of the Parachutist " DOSAAF Printing House Moscow USSR 1954

[8] Col. Charles J. Timmes "Development of Parachute and Airborne Activities in the USSR"
US Army G 2 , June 1952.

[9] Steven Zaloga "Inside the Blue Berets" Presidio Press Novato CA 1995 , page 22.

[10] "Training of the Parachutist" DOSAAF page 92.

[11] "Training of the Parachutist" DOSAAF page 108.

[12] Zaloga, page 9.

[13] "Training of the Parachutist" DOSAAF page 117

[14] Zaloga, page 22

[15] "Training of the Parachutist" DOSAAF page 115

[16] "Airborne Equipment and Airborne Defenses of the Soviet Union" Dept of the Army Washington DC 13 July 1955.

[17] FM 100-2-2 op cit.

[18] Glantz, page 90.

[19] Zaloga page 72.

[20] Gerard Devlin "Paratrooper!" St.Martins Press NY 1979, page 32.

[21] Devlin page 32.

[22] Bragg & Turner "Parachute Badges and Insignia of the World" Blandford, Dorset UK 1979 page 200

[23] Bragg & Turner, page 38

[24] Bragg & Turner, page 12

[25] Adam Jonca "Polscy Spadochroniarze" Wydawnicto Bellona, Warsaw 1994 page 21

[26] H.G.Wintersteen "Practical Flying" International Textbook Co., Scranton PA 1941, page 1.

[27] Adam Jonca "Polscy Spadochroniarze" page 10

[28] Maurice Newnham "Prelude to Glory" Sampson Low Marston Co London 1947 page 264.

[29] Maurice Newnham "Prelude to Glory", Sampson, Low & Marston Co., London, page 18

[30] John Weeks "The Airborne Soldier" Sterling Publishing NY 1982, page 82.

[31] Newnham page 15.

[32] Maurice Newnham asserts ,page 16 "Prelude to Glory", it was after 135 jumps.

[33] T.B.H. Otway "Airborne Forces" War Office, London 1951 page 403.

[34] John Weeks "Airborne Equipment" Hippocrene Books NY 1976 page 173 also "Assault from the Sky"
G.P.Putnam's Sons NY 1978 page 143.

[35] W.D. Brown "Parachutes" Sir Isaac Pitman & Sons, London 1951 page 216.

[36] Ian Wright "Bale Out!' GMS Ent. Peterborough England 1991 page 29.

[37] Dan Poynter "The Parachute Manual" Parachuting Publications Santa Barbara CA 1979 page 151.

[38] Newnham page 23.

[39] Newnham page 291.

[40] "RAF Airborne Forces Manual 1942" Arms & Armour Press, London 1979 page 39

[41] Howard P. Davies "British Parachute Forces" Arco Pub. NY 1974 pages 21-22.

[42] W.D.Brown page 187.

[43] "Parachute Training Manual" Air Ministry 1944 Chapter 1, figure 2.

[44] W.D. Brown , page 184.

[45] Wright "Bale Out!" page 29.

[46] AFEE Report no.28, section 1.2.5

[47] W.D.Brown page 216.

[48] Philip de Ste Croix "Airborne Operations" Salamander Books London 1978 page 10.

[49] John Weeks "The Airborne Soldier" page 50

[50] Newnham pages 88 – 89. The best account of how the PLF was devised.

[51] "Parachute Training Manual" chapter 1 page 12

[52] H.G.Wintersteen "Practical Flying" International Textbook Co., Scranton PA 1941 page 3.

[53] "When These US Paratroopers made Jump in '29 San Antonio Evening News 28 Sept 1944

[54] "Way of a Fighter:the memoirs of Claire Chennault" G P Putnam, NY pgs 16-17

[55] "General Billy Mitchell" Roger Burlingame McGraw Hill NY 1952 pg.8

[56] "Combat Flying Equipment 1917-1945" C.G. Sweeting Smithsonian Inst. Press Wash.D.C. pg.85

[57] Telephone conversation with MG William P Yarborough 6 Feb 1999

[58] John T. Ellis "Airborne Command and Center, Historical Section, Army Ground Forces Study 25 1946"

[59] O.J.Mink "Meet the Parachute" Reliance Mfg Co. 1944. Pages 86 & 95.

[60] C.G.Sweeting "Combat Flying Equipment. US Army Aviators Equipment 1917-1945" pg 107.

[61] Conversation between WW II rigger Ed H Thomas and Ron Brown 11 July 2000.

[62] "Establishment and Operation of a Parachute School in a Theater of Operations" The Parachute School
Fort Benning GA 26 March 1943. Pages 4 and 7

[63] Author's interview with Mr. Richard Switlik of Switlik Parachute Co., Trenton NJ 1 Oct 1998.

[64] Gerard Devlin "Paratrooper!" page 24

[65] Dan Poynter "The Parachute Manual" Santa Barbara CA 1977 page 175

[66] "Establishment and Operation of a Parachute School in a Theater of Operations" op cit page 4

[67] William Peifer "Supply by Sky: the Quartermaster Airborne Development 1950-53" Office of the Quartermaster General, US Army 1954

[68] Roger Hall "You're Stepping on My Cloak and Dagger" WW Norton NY 1957,p.112.
William J. Morgan "The OSS and I" WW Norton 1957, p. 120

[69] C.G.Sweeting , page 99.

[70] Conversation between WW II rigger Mr. Mike Cotton and Ron Brown Nov 1997

[71] Brigadier General M.D.Taylor "We also suffered a lot of casualties among our parachutists while they were trying to get out of their parachutes. We need two things immediately:

(1) quick release parachutes
(2) better anti tank weapons

Col. Charles H Coates. Notes on interviews with various infantry commanders in Normandy, France 6 June – 8 July 1944. War Dept Observer Board APO 887 5 Aug 1944.

[72] John Ellis "Airborne Command and Center Study No. 25" p.67

[73] conversation BG Ed Thomas 11 July 2000

[74] conversation between MG Yarborough and Ron Brown Feb 1998.

[75] Ellis page 65

[76] Ellis page 66

[77] Ellis page 66

[78] Devlin, page 67.

[79] Donald R. Burgett "Currahee!" Ballantine Books, New York 1967 page 39.

[80] "Establishment and Operation of a Parachute School..." page 5.

[81] "Establishment and Operation of a Parachute School..." Lecture H

[82] Flieger Abwehr Kanonen. German for anti-aircraft artillery.

[83] "Establishment and Operation of a Parachute School..." Lecture M

[84] From correspondence with Gen. Coutts to Ian B. Sutherland. 16 Feb 1998

[85] John T. Ellis "The Airborne Command and Center. Study No. 25 Historical Section" page 66.

[86] C.G. Sweeting page 107.

[87] Much of this information was extracted from "The Parachute Manual" by Dan Poynter and
"Illustrated Catalog Class 13 Supply" USAAF 30 Sept 1943, reprinted by George Petersen.

[88] "Parachute Rigger 3c and 2c" Bureau of Navy Personnel GPO 1943

[89] Ken Haney "US Marine Corps Paratroopers 1940-45" Eagle, Globe Anchor Pub. Jackson TN 1990

Acknowledgments

There are so many people to whom the author is indebted that he approaches writing this page with some trepidation, for fear that someone may be omitted. Nevertheless, honor demands this small tribute: Rex Gallogly – for paving the way and a lot of encouragement; MSG Ron Brown (USA, Ret.) – enthusiastic support, research details and interesting road trips; Tim Terrell – the loan of an X type; Malcolm Johnson – a wealth of British documents; Randy Etting – artifacts and moral support; Ian Wright – his book; and all the following who provided leads, photos, parachutes, comments, suggestions, documents, advice, and computer wrangling assistance (in no particular order):

Tim Martin
George Freeman (USA, Ret.)
Paolo Marzetti
Lt.Col. Jerry Hatzenbuhler (USA, Ret.)
James Burd
Flt.Lt. J.R.Cowsill, RAF
Michele Mischi
Col. Dave E. Severance (USMC, Ret.)
Ronald Burkett
Lt.Col. Francis C. Clagett (USMC, Ret.)
Capt. Bill Sharp ALARNG MG
William P. Yarborough (USA, Ret.)
SFC Larry Turner ALARNG
Lt.Col. Bill Morgan (USA, Ret.)
Ken Nixon
Brad Martin
Paul Crawley
Bob Johnson
Eric Queen
Larry Eads
Ken Haney
Helen Cieslik
Bob Bruce
John Morrow
Terry Carlson
Herman von Gehm
P.T. Olson
Luther Hanson
Willi Zahn
Paul Masterson
John Egger
Gregory Parsons
Joe Muchanski
Becky McCord
Kenneth Tingle (USA, Ret.)
Bill Shay
Bill Edwards
Jeff Cole
Dr. Steven Anders
Hayes Otoupalik
Albert Steiner
Gary Cochran
Brett Ousley
Gail Clark
Klaus J. Peters
Joe Keller
R. James Bender
Bob Anzuoni
Mike Jordan
Lt.Col. Ian D.W. Sutherland (USA, Ret.)
Shelton Yokomizo
Susan Hengel
Marian George
Bartek Mazur
Lewis E. Brown
Mark Jones
Jerry May
Phil Pass
Crissy Bryan
Brian Fryer
Gary Shaughnessy
Doug Bailey
Daniel Zimmerman
David List

Photo credits
Michele Mischi
Paolo Marzetti
John Morrow
John Egger
Malcolm Johnson
Tim Terrell
Ron Brown
R. James Bender
Nino Arena
Paul Crawley
Rex Gallogly
Mark Bando
Mark Jones
Shelton Yokomizo
Bob Biondi
Ron Burkett
Service Historique de L'Armee de L'Air
Sovfoto
Author's collection

NOTES

NOTES

CONCLUSION

World War II enabled theories of vertical envelopment to be tested. Despite setbacks, there were many successes such that the airborne arm was retained by all the combatant nations. Only gliders were abandoned, now replaced by helicopters. Today hardly an army exists that does not have a paratroop unit even if there is no air transport capability to lift them. Such is the value of their prestige and their prowess.